MW00835452

Vector Geometry

Gilbert de B. Robinson

Dover Publications, Inc.
Mineola, New York

Bibliographical Note

This Dover edition, first published in 2011, is an unabridged republication of the work originally published in 1962 by Allyn and Bacon, Inc., Boston.

Library of Congress Cataloging-in-Publication Data

Robinson, Gilbert de Beauregard, 1906–
Vector geometry / Gilbert de B. Robinson. — Dover ed.
p. cm.
Originally published: Boston : Allyn and Bacon, 1962.
Summary: "This brief undergraduate-level text by a prominent Cambridge-educated mathematician explores the relationship between algebra and geometry. It is the result of several years of teaching and of learning from discussion with students the most effective methods. An elementary course in plane geometry is the sole requirement, and answers to the exercises appear at the end. 1962 edition"— Provided by publisher.
Includes bibliographical references and index.
ISBN-13: 978-0-486-48160-9 (pbk.)
ISBN-10: 0-486-48160-3 (pbk.)
1. Vector analysis. 2. Geometry, Analytic. I. Title.

QA433.R63 2011
516'.182—dc22

2010052390

Manufactured in the United States by Courier Corporation
48160301
www.doverpublications.com

TO JOAN, NANCY, AND JOHN

Preface

Even though some books need none, it has become conventional to write a preface. Many people have to be thanked for their assistance in preparing the manuscript or in reading the proof—but such prefaces need not be read! Another kind of preface, however, which is sometimes not written, should be read, since it explains the sort of background which is assumed and for whom the book is intended.

BACKGROUND. No specific assumptions are made, but a student should have had a preliminary course in synthetic and also in analytical plane geometry. Permutations and combinations will come to the fore in Chapters 2 and 4, and a general feeling for algebraic processes is important throughout.

A course such as the one presented here is preliminary in Toronto to several more detailed and systematic courses in algebra and in geometry for those students who specialize in mathematics. Students who specialize in physics or in chemistry, however, may not meet these ideas again until they are brought face to face with their applications, and in such a context the practical aspects of the problem are all-important. Although a knowledge of the calculus is desirable, as the Appendix makes clear, it is not essential for understanding the ideas described here.

AIMS OF THE COURSE. This then was the problem—to give an introductory course in modern *algebra* and *geometry*—and I have proceeded on the assumption that neither is complete without the other, that they are truly two sides of the same coin.

In seeking to coordinate Euclidean, projective, and non-Euclidean geometry in an elementary way with matrices, determinants, and linear transformations, the notion of a *vector* has been exploited to the full. There is nothing new in this book, but an attempt has been made to present ideas at a level suitable to first-year students and in a manner to arouse their interest. For these associations of ideas are the stuff from which modern mathematics and many of its applications are made.

The course has been given for three successive years, and my thanks are due to three successive classes of mathematics, physics, and chemistry students who have helped me to coordinate my ideas concerning the appropriate material and the order of its presentation. Neither of these factors need be fixed and additions or alterations can easily be made, but the underlying pattern of a linear transformation and its geometrical interpretation in different contexts remains the thread which connects the different topics. The brief introduction of a quadratic transformation in Chapter 8 only serves to emphasize the pattern!

A WORD TO STUDENTS. I have tried to keep the presentation as informal as possible in an attempt to arouse and maintain interest. Some of your established ideas may be challenged in Chapter 8 but this is all part of the process! The exercises have been constructed to illustrate the subject in hand and sometimes to carry the ideas a little further, but emphasis by mere repetition has been avoided. This matter of exercises is important. You should work at them contemplatively and expect to be frustrated sometimes, for this is the only way to make the ideas your own.

G. DE B. ROBINSON

University of Toronto

References

The number of books on algebra and geometry is increasing every day, but the following list provides a reasonably diversified selection to which the reader can turn for further material.

WITH EMPHASIS ON ALGEBRA

1. Birkhoff, G., and S. MacLane, *Survey of Modern Algebra* (New York: The Macmillan Co., 1953).
2. Jaeger, A., *Introduction to Analytic Geometry and Linear Algebra* (New York: Holt, Rinehart and Winston, Inc., 1960).
3. McCoy, N. H., *Introduction to Modern Algebra* (Boston: Allyn and Bacon, Inc., 1960).
4. Murdoch, D. C., *Linear Algebra for Undergraduates* (New York: John Wiley & Sons, Inc., 1957).
5. Schreier, O., and E. Sperner, *Introduction to Modern Algebra and Matrix Theory* (New York: Chelsea Publishing Co., 1959).
6. Todhunter, I., and J. G. Leatham, *Spherical Trigonometry* (London: The Macmillan Co., 1919).
7. Thrall, R. M., and L. Tornheim, *Vector Spaces and Matrices* (New York: John Wiley & Sons, Inc., 1957).
8. Turnbull, H. W., and A. C. Aitken, *An Introduction to the Theory of Canonical Matrices* (Glasgow: Blackie and Son, Ltd., 1932).

WITH EMPHASIS ON FOUNDATIONS

1. Bonola, R., *Non-Euclidean Geometry* (New York: Dover Publications, Inc., 1955).
2. Klein, F., *Vorlesungen uber Nicht-Euklidische Geometrie* (Berlin: Springer-Verlag, 1928).
3. ———, *Elementary Mathematics from an Advanced Standpoint*, Vol. 2 (New York: The Macmillan Co., 1939).
4. Robinson, G. de B., *Foundations of Geometry* (Toronto: University of Toronto Press, 1959).

WITH EMPHASIS ON GEOMETRY

1. Askwith, E. H., *Pure Geometry* (Cambridge: Cambridge University Press, 1937).
2. Bell, R. J. T., *Coordinate Geometry of Three Dimensions* (London: The Macmillan Co., 1956).
3. Coolidge, J. L., *A Treatise on Algebraic Plane Curves* (Oxford: The Clarendon Press, 1931).
4. Coxeter, H. S. M., *The Real Projective Plane* (New York: McGraw-Hill Book Co., 1949).
5. ———, *Introduction to Geometry* (New York: John Wiley & Sons, Inc., 1961).
6. Schreier, O., and E. Sperner, *Projective Geometry* (New York: Chelsea Publishing Co., 1961).
7. Sommerville, D. M. Y., *Analytical Geometry of Three Dimensions* (Cambridge: Cambridge University Press, 1934).
8. ———, *An Introduction to the Geometry of n Dimensions* (New York: Dover Publications, Inc., 1958).

Contents

VECTOR GEOMETRY

1 LINES AND PLANES

1.1 COORDINATE GEOMETRY

The study of *geometry* is essentially the study of relations which are suggested by the world in which we live. Of course our environment suggests many relations, physical, chemical and psychological, but those which concern us here have to do with *relative positions in space* and with *distances*. We shall begin with Euclidean geometry, which is based on *Pythagoras' theorem:*

The square on the hypotenuse of a right-angled triangle is equal to the sum of the squares on the other two sides.

The statement of this fundamental result implies a knowledge of *length* and *area* as well as the notion of a *right angle*. If we know what we mean by length and may assume its invariance under what we call "motion," we can construct a right angle using a ruler and compass. We define the area of a rectangle as the product of its length and breadth. To be rigorous in these things is not desirable at this stage, but later on we shall consider a proper set of axioms for geometry.

While the Greeks did not explicitly introduce *coordinates*, it is hard to believe that they did not envisage their usefulness. The utilization of coordinates was the great contribution of Descartes in 1637, and to us now it is a most natural procedure. Take an arbitrary point O in space, the corner of the room, for instance, and three mutually perpendicular *coordinate axes*. These lines could be the three lines of intersection of the "walls" and the "floor" at O; the planes so defined we call the *coordinate planes*. In order to describe the position of a point X, we measure its perpendicular distances from each of these three planes, denoting the distances x_1, x_2, x_3 as in Figure 1.1. It is important to distinguish *direction* in making these measurements. Any point within the "room" has all its *coordinates* (x_1,x_2,x_3) positive; measurements on the opposite side of any coordinate plane would be negative. Thus the following eight combinations of sign describe the eight octants of space about O:

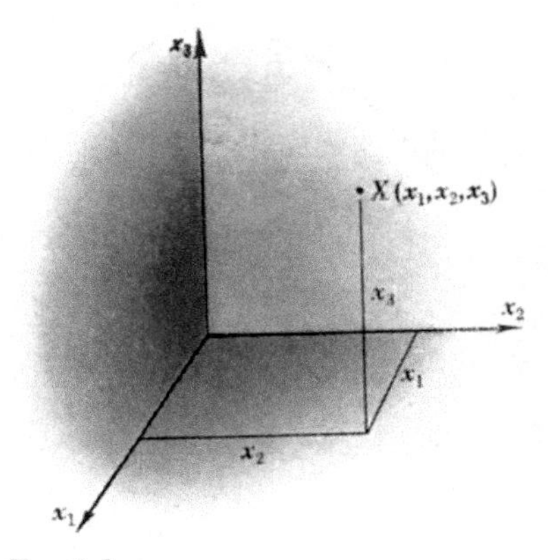

Fig. 1.1

$$+ + +, \quad + + -, \quad + - +, \quad + - -$$
$$- + +, \quad - + -, \quad - - +, \quad - - -$$

We may describe the points on the "floor" by saying that $x_3 = 0$; this is the *equation* of this coordinate plane. Limiting our attention to such points, we have plane geometry. If we call the number of mutually perpendicular coordinate axes the *dimension* of a space, then a plane has two *dimensions* and the position of each point is given by two coordinates, while *space* as we have been describing it has three dimensions.

1.2 EQUATIONS OF A LINE

If we assume that a *line* is determined uniquely by any two of its points, it is natural to seek characterizing properties dependent on these two points only. To this end we refer to Figure 1.2, assuming X to have any position on the line ZY, and complete the rectangular parallelepiped as indicated. If the coordinates of the points in question are

$$X(x_1,x_2,x_3), \quad Y(y_1,y_2,y_3), \quad Z(z_1,z_2,z_3)$$

and if XA, AB, AC are parallel to the coordinate axes with XD parallel to ZP, then from similar triangles,

$$ZX:ZA:ZB:ZC:ZD = ZY:ZP:ZQ:ZR:ZS$$

It follows from this proportionality that if we set $ZX = \tau ZY$, then

$$ZB = \tau ZQ, \quad ZC = \tau ZR, \quad ZD = \tau ZS$$

so that, in terms of coordinates,

1.21 $$x_i - z_i = \tau(y_i - z_i) \qquad i = 1,2,3$$

These equations may be rewritten thus:

1.22
$$\begin{aligned} x_1 &= z_1 + \tau(y_1 - z_1) = \tau y_1 + (1 - \tau)z_1 \\ x_2 &= z_2 + \tau(y_2 - z_2) = \tau y_2 + (1 - \tau)z_2 \\ x_3 &= z_3 + \tau(y_3 - z_3) = \tau y_3 + (1 - \tau)z_3 \end{aligned}$$

in which form they define the coordinates of X as linear functions of the *parameter* τ. Clearly, if $\tau = 0$ then $X = Z$, and if $\tau = 1$ then $X = Y$.

If we set

$$l_1 = y_1 - z_1, \quad l_2 = y_2 - z_2, \quad l_3 = y_3 - z_3$$

then l_1, l_2, l_3 are called the *direction numbers* of the line l. If X and X' are any two distinct points on l, then

$$x_i - x_i' = (\tau - \tau')(y_i - z_i) \qquad i = 1,2,3$$

so that *numbers proportional to* l_1, l_2, l_3 *are determined by any two distinct points on* l. Two lines whose direction numbers are proportional are said to be *parallel*. We can summarize these results by writing

$$(x_1 - z_1):(x_2 - z_2):(x_3 - z_3) = (y_1 - z_1):(y_2 - z_2):(y_3 - z_3) = l_1:l_2:l_3$$

It follows that we may write the *equations* of l in the symmetric form

1.231 $$\frac{x_1 - z_1}{y_1 - z_1} = \frac{x_2 - z_2}{y_2 - z_2} = \frac{x_3 - z_3}{y_3 - z_3}$$

or

1.232 $$\frac{x_1 - z_1}{l_1} = \frac{x_2 - z_2}{l_2} = \frac{x_3 - z_3}{l_3}$$

but it should be emphasized that these are valid *only if all the denominators are different from zero*, i.e., provided the line l is *not* parallel to one of the coordinate planes. As will appear in the sequel, it is the parametric equations 1.22 which are most significant. Moreover, they gener-

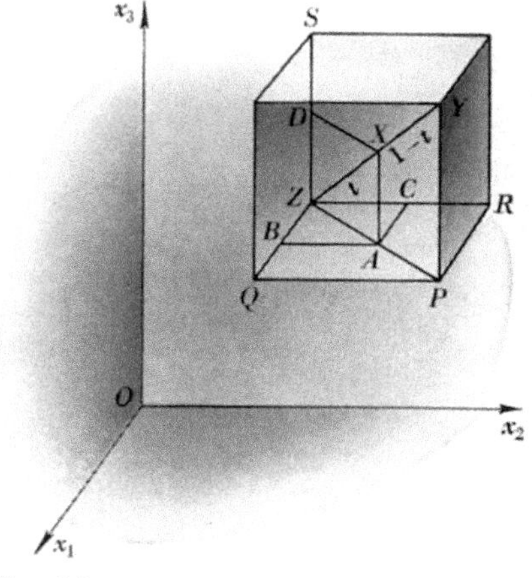

FIG. 1.2

alize easily and provide the important link between classical geometry and modern algebra.

Let us now assume that ZY makes angles $\theta_1, \theta_2, \theta_3$ with ZQ, ZR, ZS, i.e., with Ox_1, Ox_2, Ox_3. One must be careful here to insist on the direction being from Z to Y; otherwise the angles θ_i might be confused with $\pi - \theta_i$. With such a convention,

$$\begin{aligned} \cos\theta_1 &= \frac{ZQ}{ZY} = \frac{ZB}{ZX} = \lambda_1 \\ \cos\theta_2 &= \frac{ZR}{ZY} = \frac{ZC}{ZX} = \lambda_2 \qquad (1.24) \\ \cos\theta_3 &= \frac{ZS}{ZY} = \frac{ZD}{ZX} = \lambda_3 \end{aligned}$$

and $\lambda_1, \lambda_2, \lambda_3$ are called the *direction cosines* of the line l. By Pythagoras' theorem, $ZP^2 = ZQ^2 + ZR^2$, so that

$$\left(\frac{ZQ}{ZY}\right)^2 + \left(\frac{ZR}{ZY}\right)^2 + \left(\frac{ZS}{ZY}\right)^2 = \lambda_1^2 + \lambda_2^2 + \lambda_3^2 = 1 \qquad (1.25)$$

Thus, given l_1, l_2, l_3, we have

$$\lambda_1 = \frac{l_1}{\sqrt{l_1^2 + l_2^2 + l_3^2}}, \qquad \lambda_2 = \frac{l_2}{\sqrt{l_1^2 + l_2^2 + l_3^2}}, \qquad \lambda_3 = \frac{l_3}{\sqrt{l_1^2 + l_2^2 + l_3^2}} \qquad (1.26)$$

and parallel lines make equal angles with the coordinate axes.

Clearly, $\lambda_1, \lambda_2, \lambda_3$ may be substituted for l_1, l_2, l_3 in 1.232, and we may write the first set of equations of 1.22 in the form

$$\begin{aligned} x_1 &= z_1 + t\lambda_1 \\ x_2 &= z_2 + t\lambda_2 \qquad (1.27) \\ x_3 &= z_3 + t\lambda_3 \end{aligned}$$

EXERCISES

1. Find the equations, in parametric and symmetric form, of the line joining the two points $Y(1,-2,-1)$ and $Z(2,-1,0)$.

Solution. The parametric equations of the line in question are, by 1.22,

$$\begin{aligned} x_1 &= \tau + 2(1-\tau) = 2 - \tau \\ x_2 &= -2\tau - 1(1-\tau) = -1 - \tau \\ x_3 &= -\tau \end{aligned}$$

and in the symmetric form 1.231,

$$\frac{x_1 - 2}{1 - 2} = \frac{x_2 + 1}{-2 + 1} = \frac{x_3}{-1}$$

2. What are the direction cosines of the line in Exercise 1? Write the equations of the line in the form 1.27.

3. Find parametric equations for the line through the point (1,0,0) parallel to the line joining the origin to the point (0,1,2). Could these equations be written in the form 1.232?

4. Find the equations of the edges of the cube whose vertices are the eight points $(\pm 1,\pm 1,\pm 1)$, as in Figure 5 of Chapter 4.

5. Find the direction cosines of the edges of the regular tetrahedron with vertices

$$A(1,-1,-1), \quad B(-1,1,-1), \quad C(-1,-1,1), \quad D(1,1,1)$$

1.3 VECTOR ADDITION

The notion of a vector in three dimensions, or 3-space, can be introduced in two ways:

(i) A *vector* is a directed line segment of fixed length.

(ii) A *vector* $\boldsymbol{X}$ is an ordered* triple of three numbers (x_1,x_2,x_3), called the *components* of $\boldsymbol{X}$.

It is important to have both definitions clearly in mind. If we write a small arrow above the symbols to indicate *direction*, then this is determined for $\boldsymbol{V} = \overrightarrow{ZY}$ in Figure 2 by the components

$$(y_1 - z_1, \; y_2 - z_2, \; y_3 - z_3) = (v_1, v_2, v_3)$$

also, the length of $\overrightarrow{ZY}$ or the *magnitude* of $\boldsymbol{V}$ is given by

1.31 $$|\boldsymbol{V}| = \sqrt{(y_1 - z_1)^2 + (y_2 - z_2)^2 + (y_3 - z_3)^2} = \sqrt{v_1^2 + v_2^2 + v_3^2}$$

The *position* of a vector is immaterial, so we may assume it to have one end "tied" to the origin. Sometimes a vector is called "free" if it can take up any position, but this distinction is not made in either (i) or (ii). In this sense a vector is more general than any particular directed segment, and could be described as an *equivalence class*† of directed segments.

That the two definitions (i) and (ii) are equivalent follows from the theorem:

1.32 *Two vectors are equal if and only if their components are equal.*

Proof. Since the components (v_1,v_2,v_3) determine the magnitude and direction of a vector $\boldsymbol{V}$, the condition is certainly sufficient. Conversely, if $|\boldsymbol{U}| = |\boldsymbol{V}|$ then

$$u_1^2 + u_2^2 + u_3^2 = v_1^2 + v_2^2 + v_3^2$$

* 'Ordered' in the sense that the order of the components x_1, x_2, x_3 is important, so that e.g. $(x_1,x_2,x_3) \neq (x_2, x_1, x_3)$.

† Relations which are *reflexive*, *symmetric*, and *transitive* are known as *equivalence relations*, and the sets to which they apply, as *equivalence classes*. For a discussion of these ideas see Birkhoff and MacLane, *Survey of Modern Algebra*.

and if U and V have the same direction, we must have

$$u_1 = kv_1, \quad u_2 = kv_2, \quad u_3 = kv_3$$

so that $k^2 = 1$. It follows that $k = 1$ and the two vectors must coincide.

Following this line of thought, we denote the vector with components (ku_1,ku_2,ku_3) by kU so that

$$|kU| = |k||U|$$

k may be any real number, and $|k|$ is k taken positive. In particular, k may be zero, in which case kU is the *zero vector* 0 with components (0,0,0). Evidently the magnitude of 0 is zero and its direction is undefined.

We define the *sum*

$$W = U + V$$

of two vectors U and V to be the diagonal of the parallelogram formed by U and V. Alternatively, we may define W by means of the formulas

1.33 $$w_i = u_i + v_i \qquad i = 1,2,3$$

It will be sufficient to consider these definitions in the plane where a vector is defined by two components only. We take the vectors $U(u_1,u_2)$ and $V(v_1,v_2)$ and complete the parallelogram, as in Figure 1.3; it follows immediately that the components of W satisfy the relation 1.33. But there is more to be learned from the figure. For example, we arrive at the same result whether we go one way around the parallelogram or the other way around, so that

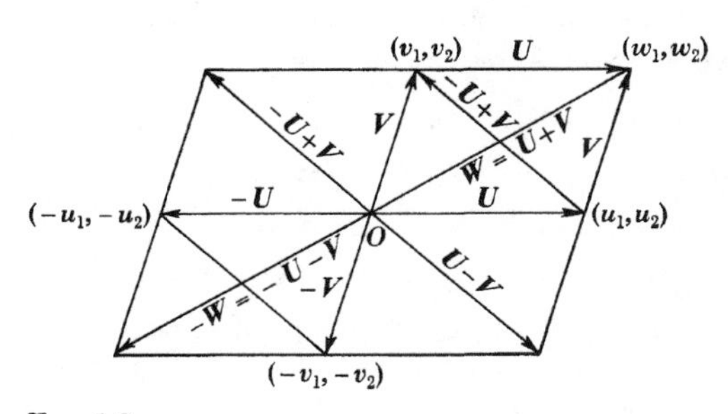

FIG. 1.3

$$U + V = V + U$$

and vector addition is *commutative*. This is also a consequence of the commutativity of addition as applied to the components in 1.33. Finally, by reversing the direction of U we obtain the vector $-U$ so that

$$-U + U = 0 = U - U$$

where 0 is the zero vector. The other diagonal of the parallelogram is the vector $-U + V$, as indicated.

Consider now the similarity between the formulas 1.33 defining vector addition and the parametric equations of a line in 1.27. If we denote by $\boldsymbol{Z}$ the vector $\overrightarrow{OZ}$ with components (z_1,z_2,z_3) and by $\boldsymbol{A}$ the vector with components $(\lambda_1,\lambda_2,\lambda_3)$, then *the relations 1.27 are the scalar equations equivalent to the vector equation*

1.34 $$\boldsymbol{X} = \boldsymbol{Z} + t\boldsymbol{A}$$

It follows that the notion of a *vector* is of central significance in Euclidean geometry. As the title of this book suggests, our purpose is to develop these ideas in several different contexts. Some of these contexts are officially "algebraic" while others are "geometric," but with this thread to guide us, we shall see their interrelations and why it is that mathematics is a living subject, changing and progressing with the introduction of new ideas.

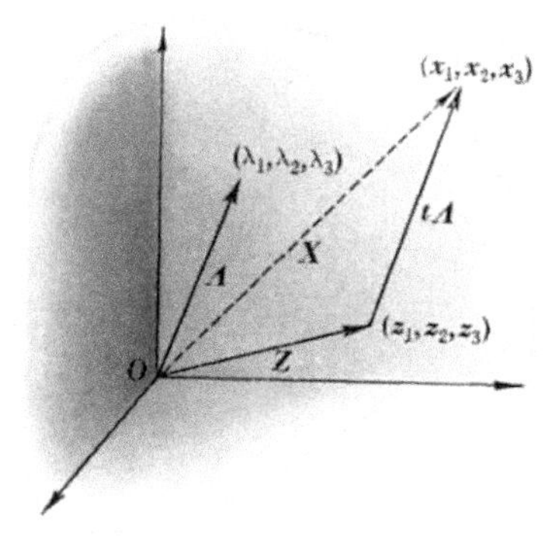

FIG. 1.4

EXERCISES

1. Show that the following vectors are equal: $\overrightarrow{OA}$, $\overrightarrow{AP}$, $\overrightarrow{QR}$ where O is the origin, A is the point $(2,-3,1)$, P is the point $(4,-6,2)$, Q is the point $(-7,3,1)$, and R is the point $(-5,0,2)$.

2. Find the components of the vectors $\overrightarrow{AB}$, $\overrightarrow{BC}$, $\overrightarrow{CA}$ where A is the point $(1,2,3)$, B is the point $(-2,3,1)$, and C is the point $(3,-2,-4)$, and show that

$$\overrightarrow{AB} + \overrightarrow{BC} + \overrightarrow{CA} = O$$

3. Determine the length and the direction cosines of the vector $\overrightarrow{AB}$ in Exercise 2. What would be the components of a parallel vector of unit length?

4. If $\boldsymbol{U}$, $\boldsymbol{V}$, $\boldsymbol{W}$ are three arbitrary vectors, show that

$$(\boldsymbol{U} + \boldsymbol{V}) + \boldsymbol{W} = \boldsymbol{U} + (\boldsymbol{V} + \boldsymbol{W})$$

(the associative law of addition).

5. Prove that the medians of any triangle ABC are concurrent.

Solution. If D is the midpoint of BC, then $\overrightarrow{OD} = \frac{1}{2}(\overrightarrow{OB} + \overrightarrow{OC})$. Since the centroid G divides AD in the ratio 2:1,

$$\overrightarrow{OG} = \tfrac{1}{3}(\overrightarrow{OA} + 2\overrightarrow{OD}) = \tfrac{1}{3}(\overrightarrow{OA} + \overrightarrow{OB} + \overrightarrow{OC})$$

Since this result is symmetric in A, B, C, the medians must be concurrent in G.

1.4 THE INNER PRODUCT

In the preceding section we defined the multiplication of a vector U by a scalar k. Such multiplication is called *scalar multiplication* and it is obviously commutative,

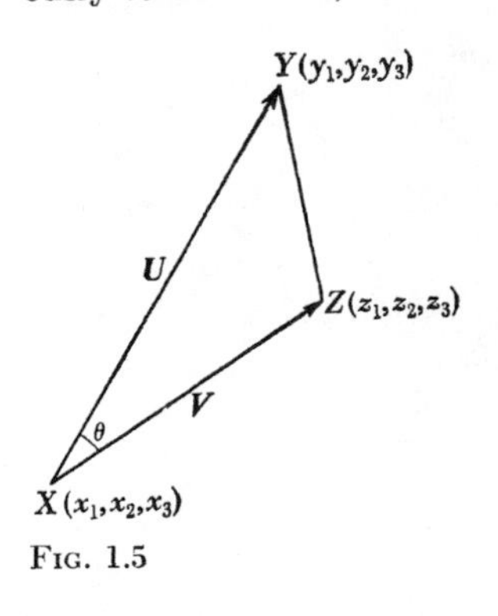

FIG. 1.5

1.41 $$kU = Uk$$

There is another kind of multiplication of vectors which is of great importance. To define it we use the generalized Pythagorean theorem to yield

1.42 $$|\overrightarrow{YZ}|^2 = |\overrightarrow{XY}|^2 + |\overrightarrow{XZ}|^2 - 2|\overrightarrow{XY}||\overrightarrow{XZ}| \cos \theta$$

so that

1.43 $$\cos \theta = \frac{|\overrightarrow{XY}|^2 + |\overrightarrow{XZ}|^2 - |\overrightarrow{YZ}|^2}{2|\overrightarrow{XY}||\overrightarrow{XZ}|}$$

Substituting from 1.31 and simplifying, we have

1.44 $$\cos \theta = \frac{(y_1 - x_1)(z_1 - x_1) + (y_2 - x_2)(z_2 - x_2) + (y_3 - x_3)(z_3 - x_3)}{|\overrightarrow{XY}||\overrightarrow{XZ}|}$$

$$= \cos \varphi_1 \cos \psi_1 + \cos \varphi_2 \cos \psi_2 + \cos \varphi_3 \cos \psi_3$$

where $\cos \varphi_1$, $\cos \varphi_2$, $\cos \varphi_3$ are the direction cosines of XY, and $\cos \psi_1$, $\cos \psi_2$, $\cos \psi_3$ are those of XZ. Since it is important to have a convenient expression for the sum of products appearing in 1.44, we define the *inner* or *scalar product* of the vectors U, V to be

1.45 $$U \cdot V = u_1v_1 + u_2v_2 + u_3v_3 = |U||V| \cos \theta$$

where U has components (u_1,u_2,u_3) and V has components (v_1,v_2,v_3).

All these formulas are valid also in the plane, but in this case a vector U has only two components (u_1,u_2), and the angles φ_1, φ_2 between U and the coordinate axes are complementary. Thus $\cos \varphi_2 = \sin \varphi_1$, so that

$$\cos \theta = \cos \varphi_1 \cos \psi_1 + \sin \varphi_1 \sin \psi_1 = \cos (\varphi_1 - \psi_1)$$

and it is convenient to write the equation of a line (note that there is now only *one* equation),

$$\frac{x_1 - z_1}{\cos \theta_1} = \frac{x_2 - z_2}{\cos \theta_2}$$

in the form

$$x_2 - z_2 = (\tan \theta_1)(x_1 - z_1)$$

Rather than try to visualize a *space* of more than three dimensions, one should think of a vector V as having n components $(v_1,v_2,\ldots,v_n)$. The sum of two vectors, $W = U + V$, is defined by the equations

$$w_i = u_i + v_i \qquad i = 1,2,\ldots n$$

and the vector kV has components $(kv_1,kv_2,\ldots kv_n)$ for any real number k.

Through use of Pythagoras' theorem, the distance between two points $Z(z_1,z_2,\ldots z_n)$ and $Y(y_1,y_2,\ldots y_n)$ is given by the relation

1.46 $$|ZY|^2 = (y_1 - z_1)^2 + (y_2 - z_2)^2 + \ldots + (y_n - z_n)^2$$

All that we have said generalizes so that

$$\cos\theta = \cos\varphi_1 \cos\psi_1 + \cos\varphi_2 \cos\psi_2 + \ldots + \cos\varphi_n \cos\psi_n$$

and

1.47 $$U\cdot V = u_1v_1 + u_2v_2 + \ldots + u_nv_n = |U||V| \cos\theta$$

In particular,

$$U\cdot U = |U|^2 = u_1^2 + u_2^2 + \ldots + u_n^2$$

and such a space is still called *Euclidean*, of n dimensions.

Finally, inner multiplication is *commutative*, and since

$$(U + V)\cdot W = U\cdot W + V\cdot W$$

it is also *distributive*. In such a relation it is not necessary to refer to the dimensionality of the space in which the vectors lie. We consider this "abstract" approach to vectors in the following section.

EXERCISES

1. Prove that if the vectors U and V are perpendicular, then $U\cdot V = 0$, and conversely.
2. Prove that each face of the regular tetrahedron in Exercise 5 of Section 1.2 is an equilateral triangle, (a) by finding the lengths of the edges and (b) by finding the angles between the edges.
3. Show that if W is perpendicular to U and also to V, then W is perpendicular to any vector $aU + bV$. How would such a vector $aU + bV$ be related to U and V? Draw a figure to illustrate the following solution.

   Solution. If $W\cdot U = 0$ and $W\cdot V = 0$, then

   $$W\cdot(aU + bV) = aW\cdot U + bW\cdot V = 0$$

   The vector $aU + bV$ would be obtained by first constructing aU collinear with U and bV collinear with V and then finding the diagonal of the parallelogram formed by aU and bV.

4. If we denote the vector $\overrightarrow{ZY}$ in Figure 5 by $\boldsymbol{W}$, then $\boldsymbol{W} = \boldsymbol{U} - \boldsymbol{V}$. Derive the relation 1.42 by calculating the inner product $\boldsymbol{W}\cdot\boldsymbol{W}$.

5. Prove the two following inequalities:

$$|\boldsymbol{U}+\boldsymbol{V}| \leqslant |\boldsymbol{U}| + |\boldsymbol{V}| \qquad \text{(triangle inequality)}$$
$$\boldsymbol{U}\cdot\boldsymbol{V} \leqslant |\boldsymbol{U}||\boldsymbol{V}| \qquad \text{(Schwarz's inequality)}$$

.5 LINEAR DEPENDENCE

As the simplest example of this important concept, let us consider a space of n dimensions and points

$$E_1 = (1,0,0,\ldots,0), \qquad E_2 = (0,1,0,\ldots,0),\ldots, \qquad E_n = (0,0,0,\ldots,1)$$

one on each coordinate axis. If we denote the vector $\overrightarrow{OE_i}$ by $\boldsymbol{E}_i$, then it is an easy extension of the ideas of the preceding section to write any vector $\mathbf{X}(x_1,x_2,\ldots x_n)$ in the form

$$\mathbf{X} = x_1\boldsymbol{E}_1 + x_2\boldsymbol{E}_2 + \ldots + x_n\boldsymbol{E}_n \tag{.51}$$

The vector $\mathbf{X}$ is said to be *linearly dependent* on the *basis* vectors $\boldsymbol{E}_i$ $(i = 1,2,\ldots n)$.

More generally, we shall say that vectors $\boldsymbol{U}, \boldsymbol{V}, \ldots \boldsymbol{W}$ are *linearly dependent* if there exists a set of constants $a, b, \ldots c$, not all zero, such that

$$a\boldsymbol{U} + b\boldsymbol{V} + \ldots + c\boldsymbol{W} = 0 \tag{.52}$$

If no such constants exist then the vectors $\boldsymbol{U}, \boldsymbol{V}, \ldots \boldsymbol{W}$ are said to be *linearly independent*.

As we have mentioned before, a vector equation 1.51 or 1.52 is equivalent to, or is a short-hand way of writing, a set of n scalar equations. For example, the scalar equations corresponding to 1.51 are

$$\begin{aligned} x_1 &= x_1\cdot 1 + x_2\cdot 0 + \ldots + x_n\cdot 0 \\ x_2 &= x_1\cdot 0 + x_2\cdot 1 + \ldots + x_n\cdot 0 \\ &\vdots \\ x_n &= x_1\cdot 0 + x_2\cdot 0 + \ldots + x_n\cdot 1 \end{aligned}$$

while those corresponding to 1.52 are

$$au_i + bv_i + \ldots + cw_i = 0 \qquad i = 1,2,\ldots n$$

We shall develop the notion of a *basis* in subsequent chapters, but the vectors $\boldsymbol{E}_i$ are particularly important; not only are they pairwise *orthogonal*, i.e., perpendicular, but they are also *normal*, i.e., of unit length. We express both these facts by writing

1.53
$$E_i \cdot E_j = \begin{cases} 0, & i \neq j \\ 1, & i = j \end{cases}$$

In 3-space it is sometimes convenient to use the notation $x_1 = x$, $x_2 = y$, $x_3 = z$, in which case we write $E_1 = i$, $E_2 = j$, $E_3 = k$. The advantage of the suffix notation, however, is that it extends to any number of dimensions.

It is interesting to see that we could have approached our subject from a purely abstract point of view, defining an *abstract vector space* $\mathcal{V}$ as a set of vectors $A, B, C, \ldots$ with the property of addition such that:

(i) If A and B are vectors in $\mathcal{V}$ so also is $A + B$

(ii) $A + B = B + A$ (commutative law of addition)

(iii) $(A + B) + C = A + (B + C)$ (associative law of addition)

(iv) There exists a vector in $\mathcal{V}$ called the *zero* vector 0 such that $A + 0 = A = 0 + A$

(v) With every vector A in $\mathcal{V}$ is associated a vector $-A$ such that

$$A + (-A) = 0 = (-A) + A$$

(vi) With every real number k and vector A in $\mathcal{V}$ is associated a vector $kA = Ak$ such that $k_1A + k_2A = (k_1 + k_2)A$ and $k_1(k_2A) = k_1k_2A$. We assume that $1A = A$ for all A.

1.54 *Definition* The number of linearly independent vectors in $\mathcal{V}$ is called the *dimension* of $\mathcal{V}$.

We may introduce the notion of an inner product $A \cdot B$ by assuming this operation to satisfy the further axioms

(vii) $A \cdot B = B \cdot A$ (commutative law of inner multiplication)

(viii) $A \cdot (B + C) = A \cdot B + A \cdot C$ (distributive law)

(ix) $(kA \cdot B) = (A \cdot kB) = k(A \cdot B)$

(x) For any vector A in $\mathcal{V}$, $A \cdot A$ is a real positive number or zero.

(xi) $A \cdot A = 0$ implies that $A = 0$.

Thus we may set $|A|^2 = A \cdot A$ and call $|A| \geqslant 0$ the *magnitude* of A. Similarly, for any two vectors A, B we may define

$$\cos \theta = \frac{A \cdot B}{|A||B|}$$

thus avoiding the use of components at all. However, the geometrical definitions given in Section 3 provide the most familiar realization of a vector space $\mathcal{V}$ and the only one with which we shall be concerned.

EXERCISES

1. If A and B have coordinates $(-1,2,0)$ and $(2,1,-1)$ respectively, express the vector $\overrightarrow{AB}$ in terms of the basis vectors i, j, k.

2. Find the lengths of the two diagonals of the parallelogram formed by the vectors $\overrightarrow{OA}$ and $\overrightarrow{OB}$ in Exercise 1.

3. Prove that the two vectors $\boldsymbol{U} = \boldsymbol{i} - 8\boldsymbol{j} + 2\boldsymbol{k}$ and $\boldsymbol{V} = 6\boldsymbol{i} + 2\boldsymbol{j} + 5\boldsymbol{k}$ are orthogonal, and find $|\boldsymbol{U}|$ and $|\boldsymbol{V}|$.

4. Find $\mathbf{X}$ perpendicular to $3\boldsymbol{i} - \boldsymbol{j} + 2\boldsymbol{k}$ and $2\boldsymbol{i} + 5\boldsymbol{j} + 7\boldsymbol{k}$, and such that $|\mathbf{X}| = 2$.

5. In the regular tetrahedron with vertices $A(1,-1,-1)$, $B(-1,1,-1)$, $C(-1,-1,1)$, $D(1,1,1)$, prove that the vectors $\overrightarrow{AB}$, $\overrightarrow{AC}$, $\overrightarrow{AD}$ are linearly independent.

6. Express the vector $\overrightarrow{BC}$ in Exercise 5 as a linear combination of the vectors $\overrightarrow{AB}$ and $\overrightarrow{AC}$. Express each of these vectors in terms of the basis vectors $\boldsymbol{i}, \boldsymbol{j}, \boldsymbol{k}$, and show that the same relation holds.

   Solution. The components of the vectors in question are

$$\overrightarrow{AB}(-2,2,0), \quad \overrightarrow{BC}(0,-2,2), \quad \overrightarrow{AC}(-2,0,2)$$

   so that $\overrightarrow{BC} = \overrightarrow{AC} - \overrightarrow{AB}$. In terms of $\boldsymbol{i}, \boldsymbol{j}, \boldsymbol{k}$,

$$\overrightarrow{AB} = -2\boldsymbol{i} + 2\boldsymbol{j}, \quad \overrightarrow{BC} = -2\boldsymbol{j} + 2\boldsymbol{k}, \quad \overrightarrow{AC} = -2\boldsymbol{i} + 2\boldsymbol{k}$$

   and clearly the same relation holds good.

7. If the centroid G of the tetrahedron in Exercise 5 is defined by the equation

$$\overrightarrow{OG} = \tfrac{1}{4}(\overrightarrow{OA} + \overrightarrow{OB} + \overrightarrow{OC} + \overrightarrow{OD})$$

   find the coordinates of G.

8. Prove that the lines joining the midpoints of opposite edges of any tetrahedron concur in the centroid of the tetrahedron.

9. Prove that the lines joining the vertices to the centroids of opposite faces concur in the centroid of the tetrahedron.

1.6 EQUATIONS OF A PLANE

Let us begin by writing the parametric equations 1.21 or 1.22 of a line in the vector form

1.61
$$(\mathbf{X} - \mathbf{Z}) - \tau(\mathbf{Y} - \mathbf{Z}) = 0$$

or

1.62
$$\mathbf{X} = \tau\mathbf{Y} + (1 - \tau)\mathbf{Z}$$

1.63 *The vectors* $\mathbf{X} - \mathbf{Z}$ *and* $\mathbf{Y} - \mathbf{Z}$ *are linearly dependent if and only if the points* X, Y, Z *are collinear.*

If now X, Y, Z are *not* collinear, they will define a plane π; and if U is any point in π we may complete the parallelogram as in Figure 1.6 and write

1.64 $$(\mathbf{U} - \mathbf{Z}) - \lambda(\mathbf{X} - \mathbf{Z}) - \mu(\mathbf{Y} - \mathbf{Z}) = \mathbf{0}$$

or

1.65 $$\mathbf{U} = \lambda \mathbf{X} + \mu \mathbf{Y} + (1 - \lambda - \mu)\mathbf{Z}$$

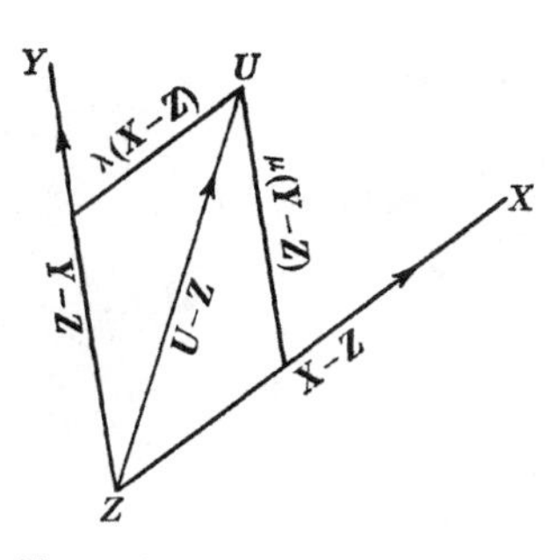

FIG. 1.6

Essentially, *we have established a coordinate system in* π *with origin* Z *and axes* ZX *and* ZY. The equation of ZX is $\mu = 0$ while that of ZY is $\lambda = 0$, and the *coordinates* of U are (λ,μ). Note that these axes ZX and ZY need not be orthogonal; all that we do require is that λ, μ be determined by lines *parallel* to ZX and ZY. Since all these steps are reversible, we conclude that

1.66 *The vectors* $\mathbf{U} - \mathbf{Z}$, $\mathbf{X} - \mathbf{Z}$, *and* $\mathbf{Y} - \mathbf{Z}$ *are linearly dependent if and only if the points* U, X, Y, Z *are coplanar.*

However, we can approach the problem from quite a different point of view. Let us assume that π passes through the point $Z(z_1,z_2,z_3)$ and is perpendicular to a given vector $\mathbf{U}$. Then if X is any point of π, the vector $\mathbf{V} = \overrightarrow{ZX}$ is perpendicular to $\mathbf{U}$ so that $\mathbf{U} \cdot \mathbf{V} = 0$, or

1.67 $$u_1(x_1 - z_1) + u_2(x_2 - z_2) + u_3(x_3 - z_3) = 0$$

and this is the *equation* of π. Note that the normal vector $\mathbf{U}$ is not unique since 1.67 may be multiplied through by any constant $k \neq 0$.

We conclude this discussion of planes in space by computing the perpendicular distance p from a point $X(x_1,x_2,x_3)$ of general position to the plane π with equation 1.67. If this distance is measured along a normal $\overrightarrow{XN}$ making an angle θ with $\overrightarrow{XZ}$, then

$$p = |\overrightarrow{XZ}| \cos \theta$$

$$= \frac{|\overrightarrow{XZ}||\mathbf{U} \cdot (\mathbf{X} - \mathbf{Z})|}{|\overrightarrow{XZ}||\mathbf{U}|} \qquad \text{by 1.45}$$

$$= \frac{|u_1(x_1 - z_1) + u_2(x_2 - z_2) + u_3(x_3 - z_3)|}{\sqrt{u_1^2 + u_2^2 + u_3^2}}$$

If the equation 1.67 had been simplified and written in the form

$$u_1x_1 + u_2x_2 + u_3x_3 + u_4 = 0$$

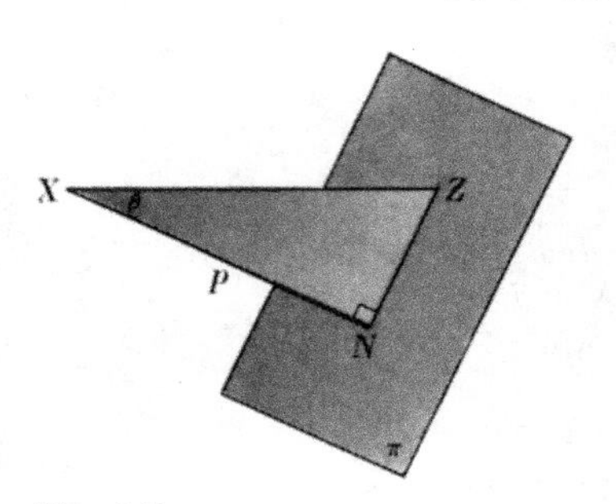

FIG. 1.7

then we would have

$$\textit{1.68} \qquad p = \frac{|u_1x_1 + u_2x_2 + u_3x_3 + u_4|}{\sqrt{u_1^2 + u_2^2 + u_3^2}}$$

The important question: *What locus is represented by the general linear equation with real coefficients?* can be answered by first finding a point $Z(z_1,z_2,z_3)$ whose coordinates satisfy the equation, and then rearranging it in the form 1.67. We conclude from this and Section 1.2 that:

1.69 *Every linear equation in x_1, x_2, x_3 represents a plane; two linear equations represent the line of intersection of the two planes, unless the two planes are parallel.*

The determination of the parametric or symmetric form of the equations of a line l, when l is defined by two linear equations, is illustrated by the following example.

Example. Consider the intersection l of the two planes

$$\begin{aligned} x_1 - x_2 + 2x_3 &= 1 \\ x_1 + x_2 - x_3 &= 3 \end{aligned}$$

In order to find a point on l we first look for the intersection of l with, say, the plane $x_3 = 0$, which yields the point $Z(2,1,0)$. The direction numbers l_1, l_2, l_3 of l must satisfy the two linear equations

$$\begin{aligned} l_1 - l_2 + 2l_3 &= 0 \\ l_1 + l_2 - l_3 &= 0 \end{aligned}$$

which express the fact that l is orthogonal to the normal direction of each plane containing it. Thus

$$l_1:l_2:l_3 = -1:3:2$$

so that the parametric equations of l may be written

$$x_1 = 2 - t, \quad x_2 = 1 + 3t, \quad x_3 = 2t$$

and the symmetric equations

$$\frac{x_1 - 2}{-1} = \frac{x_2 - 1}{3} = \frac{x_3}{2}$$

The direction cosines of l are $-1/\sqrt{14}$, $3/\sqrt{14}$, $2/\sqrt{14}$.

EXERCISES

1. Find the equations of the faces of the tetrahedron with vertices

$$A(1,1,1), \quad B(1,-1,-1), \quad C(-1,1,-1), \quad D(-1,-1,1)$$

and their angles of intersection.

Solution. The equation of the face ABC may be taken to be

$$u_1(x_1 - 1) + u_2(x_2 - 1) + u_3(x_3 - 1) = 0$$

Substituting the coordinates of B and C,

$$\begin{aligned} -2u_2 - 2u_3 &= 0 \\ -2u_1 \qquad - 2u_3 &= 0 \end{aligned}$$

so that the equation of ABC is

$$(x_1 - 1) + (x_2 - 1) - (x_3 - 1) = 0$$

Similarly, the equation of the face ABD is

$$(x_1 - 1) - (x_2 - 1) + (x_3 - 1) = 0, \text{ etc.}$$

The angle θ between the faces will be the angle between their normals, properly directed, so that

$$\cos\theta = \frac{-1.1 + 1.1 + 1.1}{3} = \frac{1}{3}$$

2. Find the equations of the faces of the octahedron with vertices

$$\begin{array}{lll} A(1,0,0), & B(0,1,0), & C(0,0,1) \\ A'(-1,0,0), & B'(0,-1,0), & C'(0,0,-1) \end{array}$$

and the angles between faces which (a) intersect in an edge, (b) intersect in a vertex. Which faces are parallel to one another?

3. (a) Give the components of vectors parallel to each of the edges of the octahedron in Exercise 2.
(b) Express each of these vectors in terms of the basis vectors E_1, E_2, E_3.

4. Show that the plane $u_1x_1 + u_2x_2 + u_3x_3 + u_4 = 0$
(i) meets Ox_i at a distance $-u_4/u_i (i = 1,2,3)$ from 0
(ii) is parallel to Ox_i if $u_i = 0$
(iii) is perpendicular to Ox_i if $u_j = u_k = 0 (j \neq k \neq i)$
Generalize these statements to an Euclidean space of n dimensions.

2 DETERMINANTS AND LINEAR EQUATIONS

2.1 THE PROBLEM DEFINED

By introducing coordinates, Descartes aimed to make it possible to solve geometrical problems "analytically." Thus, as we saw in Chapter 1, the study of lines and planes in space is translated into the study of simultaneous linear equations. There are two important aspects of this problem which become confused when the number of variables is small, namely, (a) the finding of *actual solutions* of a given system of simultaneous equations, and (b) the investigation of the *properties* of such solutions in general without explicitly determining of them. Though we shall introduce the abbreviation known as a *determinant* for a "multilinear" polynomial expression and use determinants to solve systems of linear equations, the reader should be warned that the real importance of determinants is theoretical rather than practical.

Let us begin with the simple case

2.111 $$a_{11}x_1 + a_{12}x_2 = a_{10}$$

2.112 $$a_{21}x_1 + a_{22}x_2 = a_{20}$$

If we multiply 2.111 by a_{22} and 2.112 by $-a_{12}$ and add, we have

2.113 $$(a_{11}a_{22} - a_{21}a_{12})x_1 = (a_{10}a_{22} - a_{20}a_{12})$$

Similarly, if we multiply 2.111 by $-a_{21}$ and 2.112 by a_{11} and add, we have

2.114 $$(a_{11}a_{22} - a_{21}a_{12})x_2 = (a_{11}a_{20} - a_{21}a_{10})$$

For convenience in writing the solution we set

2.12 $$a_{11}a_{22} - a_{21}a_{12} = \begin{vmatrix} a_{11} & a_{12} \\ a_{21} & a_{22} \end{vmatrix} = \Delta$$

called a *determinant* of order 2, so that

$$\Delta x_1 = \begin{vmatrix} a_{10} & a_{12} \\ a_{20} & a_{22} \end{vmatrix} = \Delta_1, \qquad \Delta x_2 = \begin{vmatrix} a_{11} & a_{10} \\ a_{21} & a_{20} \end{vmatrix} = \Delta_2$$

Observe that the determinant Δ_1 (Δ_2) is formed by replacing the first (second) column of Δ by the vector (a_{10},a_{20}). Geometrically, we have found the point of intersection of two coplanar lines when $\Delta \neq 0$. If $\Delta = 0$, the two lines are parallel and so have no point of intersection unless they coincide.

Example. The two equations

$$x_1 - x_2 = 1$$
$$x_1 + x_2 = 3$$

have as solution

$$x_1 = \frac{\begin{vmatrix} 1 & -1 \\ 3 & 1 \end{vmatrix}}{\begin{vmatrix} 1 & -1 \\ 1 & 1 \end{vmatrix}} = \frac{4}{2} = 2, \qquad x_2 = \frac{\begin{vmatrix} 1 & 1 \\ 1 & 3 \end{vmatrix}}{\begin{vmatrix} 1 & -1 \\ 1 & 1 \end{vmatrix}} = \frac{2}{2} = 1$$

The two equations

$$x_1 - x_2 = 1$$
$$2x_1 - 2x_2 = 1$$

represent parallel lines since $\Delta = 0$; they have no solution and are said to be *inconsistent.*

Let us now seek the solution of the three linear equations

2.131 $$a_{11}x_1 + a_{12}x_2 + a_{13}x_3 = a_{10}$$

2.132 $$a_{21}x_1 + a_{22}x_2 + a_{23}x_3 = a_{20}$$

2.133 $$a_{31}x_1 + a_{32}x_2 + a_{33}x_3 = a_{30}$$

To proceed systematically, we begin by reducing the problem to the case $n = 2$ by eliminating x_3, thus:

2.134 $$(a_{11}a_{23} - a_{21}a_{13})x_1 + (a_{12}a_{23} - a_{22}a_{13})x_2 = a_{10}a_{23} - a_{20}a_{13}$$

$$(a_{11}a_{33} - a_{31}a_{13})x_1 + (a_{12}a_{33} - a_{32}a_{13})x_2 = a_{10}a_{33} - a_{30}a_{13} \tag{2.135}$$

$$(a_{21}a_{33} - a_{31}a_{23})x_1 + (a_{22}a_{33} - a_{32}a_{23})x_2 = a_{20}a_{33} - a_{30}a_{23} \tag{2.136}$$

If we multiply the equation 2.134 by a_{33}, 2.135 by a_{23}, and subtract, we obtain 2.136 multiplied by a_{13}, but for the sake of symmetry we retain all three equations. If now we eliminate x_2 by multiplying 2.134 by $-a_{32}$, 2.135 by a_{22}, 2.136 by $-a_{12}$, and add, we arrive at the equation

$$\begin{aligned}&[a_{11}(a_{22}a_{33} - a_{32}a_{23}) - a_{21}(a_{12}a_{33} - a_{32}a_{13}) + a_{31}(a_{12}a_{23} - a_{22}a_{13})]x_1 \\ &\quad = a_{10}(a_{22}a_{33} - a_{32}a_{23}) - a_{20}(a_{12}a_{33} - a_{32}a_{13}) + a_{30}(a_{12}a_{23} - a_{22}a_{13})\end{aligned} \tag{2.14}$$

By a similar procedure we could obtain equations for x_2 and x_3.

Again, let us set

$$a_{11}(a_{22}a_{33} - a_{32}a_{23}) - a_{21}(a_{12}a_{33} - a_{32}a_{13}) + a_{31}(a_{12}a_{23} - a_{22}a_{13}) = \begin{vmatrix} a_{11} & a_{12} & a_{13} \\ a_{21} & a_{22} & a_{23} \\ a_{31} & a_{32} & a_{33} \end{vmatrix} = \Delta \tag{2.15}$$

so that

$$\Delta x_1 = \Delta_1, \quad \Delta x_2 = \Delta_2, \quad \Delta x_3 = \Delta_3$$

As before, Δ_i is obtained by replacing the ith column of Δ by the vector (a_{10},a_{20},a_{30}).

It is important to observe that *the groups of terms multiplying* a_{11}, a_{21}, a_{31} *in 2.15 are just the second-order determinants obtained by crossing out the row and column containing* a_{11}, a_{21}, a_{31} *in* Δ *multiplied by* ± 1. In fact, we could multiply 2.131 by

$$(a_{22}a_{33} - a_{32}a_{23}) = A_{11}$$

2.132 by

$$-(a_{12}a_{33} - a_{32}a_{13}) = A_{21}$$

2.133 by

$$(a_{12}a_{23} - a_{22}a_{13}) = A_{31}$$

and add to obtain 2.14 directly, since

$$a_{11}A_{11} + a_{21}A_{21} + a_{31}A_{31} = \Delta \tag{2.161}$$

$$a_{1i}A_{11} + a_{2i}A_{21} + a_{3i}A_{31} = 0 \qquad i = 2,3 \tag{2.162}$$

$$a_{10}A_{1j} + a_{20}A_{2j} + a_{30}A_{3j} = \Delta_j \qquad j = 1,2,3 \tag{2.163}$$

The equation 2.161 leads to an *inductive* definition of a determinant of order n which runs as follows:

2.17 Definition If $A = (a_{ij})$ is an $n \times n$ array or *matrix*, the *minor* M_{ij} of an element a_{ij} in $|A|$ is the determinant of that $(n-1) \times (n-1)$ matrix obtained from A by striking out the ith row and jth column, and the *co-factor* A_{ij} is defined by the equation

2.171 $$A_{ij} = (-1)^{i+j}M_{ij} \qquad i,j = 1,2, \ldots n$$

We define the determinant $|A|$ of A by the relation

2.172 $$|A| = a_{11}A_{11} + a_{21}A_{21} + \ldots + a_{n1}A_{n1}$$

Since $|A|$ has been defined in 2.12 for $n = 2$ and in 2.15 for $n = 3$, the inductive definition is complete.

We summarize the method developed above, expressing the solution of a system of n linear equations in n unknowns in

2.18 Cramer's Rule *The solution of n linear equations,*

2.181
$$\begin{aligned} a_{11}x_1 + a_{12}x_2 + \ldots + a_{1n}x_n &= a_{10} \\ a_{21}x_1 + a_{22}x_2 + \ldots + a_{2n}x_n &= a_{20} \\ &\;\vdots \\ a_{n1}x_1 + a_{n2}x_2 + \ldots + a_{nn}x_n &= a_{n0} \end{aligned}$$

is given by the formulas

2.182 $$\Delta x_1 = \Delta_1, \quad \Delta x_2 = \Delta_2, \ldots, \quad \Delta x_n = \Delta_n$$

where $\Delta = |a_{ij}| \neq 0$, *and* Δ_i *is obtained by replacing the ith column of* Δ *by the vector*

$$(a_{10}, a_{20}, \ldots, a_{n0})$$

As in the case $n = 2$, we may interpret these results geometrically, but we postpone this until we have developed the general properties of determinants suggested above, proving Cramer's rule in Section 2.3.

EXERCISES

1. Evaluate the following determinants by multiplying out all the terms in their "expansions" according to 2.15 above:

$$\begin{vmatrix} 1 & 0 & 1 \\ 0 & 3 & -5 \\ -2 & 1 & 4 \end{vmatrix}, \quad \begin{vmatrix} 5 & -3 & 1 \\ -4 & 1 & -1 \\ 3 & 5 & -1 \end{vmatrix}, \quad \begin{vmatrix} 2 & -3 & 1 \\ -6 & 1 & 5 \\ 3 & 2 & -5 \end{vmatrix}$$

2. Solve the simultaneous equations

$$x_1 + x_3 = 0, \qquad 3x_2 - 5x_3 = 1, \qquad 2x_1 - x_2 - 4x_3 = 2$$

(a) by elimination and (b) by determinants.

3. Have the equations

$$2x_1 - 3x_2 + x_3 = 0, \qquad 6x_1 - x_2 - 5x_3 = 1$$
$$3x_1 + 2x_2 - 5x_3 = 3$$

a solution? If not, show explicitly that they are inconsistent.

2.2 DETERMINANTS

There is another definition of a determinant which is important:

2.21 Definition If $(a_{ij}) = A$ is an $n \times n$ "array" or *matrix*, then the determinant of A is the expression

2.211
$$|A| = \sum_{n!} \operatorname{sgn} \pi \cdot a_{\pi(1),1} \cdot a_{\pi(2),2} \cdot \cdots \cdot a_{\pi(n),n}$$

where $\pi(1), \pi(2), \ldots, \pi(n)$ is a permutation of $1,2,\ldots,n$, $\operatorname{sgn} \pi = \pm 1$ according as π is *even* or *odd*, and the summation is over all $n!$ such permutations.

Before we attempt to reconcile these two definitions we must make clear the notion of the *evenness* or *oddness* of a permutation. To this end consider all $3!$ arrangements of the symbols 1,2,3:

2.22 123, 132, 213, 231, 312, 321

By comparison with the initial arrangement 123, we may construct for each arrangement the *permutation* π which accomplishes the change, by starting with any given symbol a and setting next on the right the symbol b into which a is transformed by π, then setting next to $b = \pi(a)$ the symbol $c = \pi(b)$, and so on. In due course, we return to a, completing the *cycle*. If this exhausts the symbols, we have the desired permutation π. If not, we begin again with another symbol not already considered and so construct all the cycles of π. If we do this for the arrangements of 2.22 we obtain

2.23 (1)(2)(3), (1)(23), (12)(3), (123), (132), (13)(2)

or

2.24 I, (23), (12), (123), (132), (13)

Cycles of length one are usually omitted and the identity permutation is represented by I.

It is often convenient to write a permutation π in *two-rowed* form, where $\pi(a)$ is placed immediately beneath a. Thus, the permutations 2.24 could also be written

$$\begin{pmatrix}123\\123\end{pmatrix}, \quad \begin{pmatrix}123\\132\end{pmatrix}, \quad \begin{pmatrix}123\\213\end{pmatrix}, \quad \begin{pmatrix}123\\231\end{pmatrix}, \quad \begin{pmatrix}123\\312\end{pmatrix}, \quad \begin{pmatrix}123\\321\end{pmatrix}$$

It is important that the order of writing the various columns of the two-rowed form does not matter, so that permutations can be combined as in the following paragraph.

Observe that the permutations (123) and (132) can be written as a sequence of *transpositions*, i.e., cycles of length 2, in the following manner:

$$(123) = \begin{pmatrix}123\\132\\231\end{pmatrix} = (12)(23), \qquad (132) = \begin{pmatrix}123\\132\\312\end{pmatrix} = (13)(32)$$

since it is customary to operate or *multiply* permutations from *right* to *left*. It can easily be verified that

$$(123) = (12)(23) = (23)(13) = (23)(13)tt$$

for any transposition t, since $tt = I$. Thus the number of transpositions in terms of which (123) may be expressed seems to preserve its parity, though the transpositions themselves may differ; certainly the number of such transpositions is not unique.

2.25 *Definition* A permutation is said to be *even* or *odd* according as the number of transpositions required to express it is even or odd.

If this definition is to be significant we must prove the following

2.26 Theorem *The number of transpositions in terms of which a given permutation π on n symbols may be expressed is always even or always odd.*

Proof. Consider the general case and suppose that the permutation π operates on the *subscripts* of $x_1, x_2, \ldots x_n$. To prove the theorem we construct the function

$$\mathrm{P} = \prod_{i<j} (x_i - x_j)$$

and consider the effect of π on P. If $\pi = (ij)$:

2.261

$$\begin{array}{lll} (x_i - x_j) \rightarrow -(x_i - x_j) & & \\ (x_k - x_i) \leftrightarrow (x_k - x_j) & \text{for} & k < i < j \\ \left.\begin{array}{l}(x_i - x_k) \rightarrow -(x_k - x_j) \\ (x_k - x_j) \rightarrow -(x_i - x_k)\end{array}\right\} & \text{for} & i < k < j \\ (x_i - x_k) \leftrightarrow (x_j - x_k) & \text{for} & i < j < k \end{array}$$

Since these are the only factors of P affected by $\pi = (ij)$, we conclude that $\pi(\mathrm{P}) = -\mathrm{P}$ *for any transposition* π.

On the other hand, $\pi(\mathrm{P})$ is well defined for any permutation π, so $\pi(\mathrm{P})$ is either P or $-$P. We conclude that π must always be expressible as a product of an *even* or *odd* number of transpositions, as required.

With these explanations, our definition 2.21 of a determinant is complete, and it agrees with that of the preceding section for $n = 2,3$. (Note that I, (123), (132) are *even* while (12), (13), (23) are *odd* permutations, yielding the signs of 2.15 as written.) In order to identify the two definitions we observe that:

(i) No term in the expansion of $|A|$ in 2.21 has two factors a_{ij} with the same first or second suffixes, so that $|A|$ is *linear* in the elements of any row or column.

(ii) Since the $\pi(1), \pi(2), \ldots \pi(n)$ are just $1,2, \ldots n$ rearranged, we could equally well suppose the *first* suffixes to be arranged in natural order $1,2, \ldots n$ and the second suffixes permuted by the inverse permutation π^{-1} of π; for example, if

$$\pi = (123) = \binom{123}{231}$$

then turned upside down,

$$\binom{123}{312} = (132) = \pi^{-1}$$

Clearly $\pi\pi^{-1} = I = \pi^{-1}\pi$, and π^{-1} is even or odd when π is even or odd.

From (ii) we conclude that

2.27 $$|A| = |A^t|$$

where A^t is the *transpose* of the matrix A, obtained by writing the ith row of A as the ith column of A^t.

The property (i) of $|A|$ is more subtle. Certainly we can collect together those terms in 2.211 which include a_{11} as a factor; the number of these is $(n - 1)!$ and they are just those terms which make up $M_{11} = A_{11}$ as defined in the preceding section. Similarly, we may collect those terms in 2.211 which include a_{21} as a factor; they will be distinct from those which include a_{11} by (i) and will make up $-M_{21}$. To prove this last statement it is sufficient to observe that, to obtain the π's corresponding to terms containing a_{21} as a factor, we need only operate on those π's corresponding to terms containing a_{11} as a factor by the transposition (12), and this accounts for the minus sign. By 2.171,

$$-M_{21} = A_{21}$$

Again, we collect the terms containing a_{31} as a factor; they are distinct by (i) and make up $M_{31} = A_{31}$, and so on. We conclude that

2.28 $$|A| = a_{11}A_{11} + a_{21}A_{21} + \ldots + a_{n1}A_{n1}$$

as before.

The argument of the preceding paragraph is quite general, and by introducing an extra transposition we can shift the first column into the second column position or, by a further transposition, into the third column position, and so on, leaving the minors of the elements in the column unchanged. Otherwise described, one can think of moving a_{ij} into the position of a_{11}, leaving its minor unchanged. Since this same process can be applied to both the rows of A and those of A^t, i.e., the columns of A, the number of transpositions is

$$(i - 1) + (j - 1) = i + j - 2$$

and since $(-1)^{i+j} = (-1)^{i+j-2}$, we have explained the significance of 2.171. Moreover, we have generalized 2.28 so that we have the following important result:

2.29 $$|A| = \sum_{i=1}^{n} a_{ik}A_{ik} = \sum_{j=1}^{n} a_{kj}A_{kj} \qquad 1 \leqslant k \leqslant n$$

2.3 EVALUATION OF A DETERMINANT

While the definition 2.21 of $|A|$ is theoretically significant, $n!$ therein increases so rapidly that to evaluate an $n \times n$ determinant by calculating each term in its expansion becomes prohibitive. Nor is the calculation of the $(n-1) \times (n-1)$ minors more feasible, so we must develop a third and more practical method of evaluating $|A|$. To this end we prove a sequence of theorems of disarming simplicity.

2.31 *Interchanging any two columns (rows) of (A) changes $|A|$ into $-|A|$.*

Proof. Since such an interchange corresponds to introducing an extra transposition into each π in 2.211, so changing the sign of each term in the expansion of $|A|$ or $|A^t| = |A|$, and the statement follows.

2.32 *If two columns (rows) of A are the same, $|A| = 0$.*

Proof. Suppose the ith column of A is equal to the jth column. Introducing the extra transposition (ij) does not affect A, but $|A|$ changes sign. Hence $|A| = -|A| = 0$, and similarly for A^t.

2.33 *Multiplication of a column (row) of A by k changes $|A|$ into $k|A|$.*

Proof. This follows immediately from 2.211, since every term in the expansion contains just one factor from each row and each column.

2.34 *Adding a constant multiple of a column (row) of A to another column (row) of A leaves $|A|$ unchanged.*

Proof. Let us assume we are adding k times the ith column to the jth column of A, which yields:

$$\begin{vmatrix} a_{11} \cdot\cdot a_{1i} \cdot\cdot a_{1j} + ka_{1i} \cdot\cdot a_{1n} \\ a_{21} \cdot\cdot a_{2i} \cdot\cdot a_{2j} + ka_{2i} \cdot\cdot a_{2n} \\ \cdot \\ \cdot \\ \cdot \\ a_{n1} \cdot\cdot a_{ni} \cdot\cdot a_{nj} + ka_{ni} \cdot\cdot a_{nn} \end{vmatrix}$$

$$= \begin{vmatrix} a_{11} \cdot\cdot a_{1i} \cdot\cdot a_{1j} \cdot\cdot a_{1n} \\ a_{21} \cdot\cdot a_{2i} \cdot\cdot a_{2j} \cdot\cdot a_{2n} \\ \cdot \\ \cdot \\ \cdot \\ a_{n1} \cdot\cdot a_{ni} \cdot\cdot a_{nj} \cdot\cdot a_{nn} \end{vmatrix} + k \begin{vmatrix} a_{11} \cdot\cdot a_{1i} \cdot\cdot a_{1i} \cdot\cdot a_{1n} \\ a_{21} \cdot\cdot a_{2i} \cdot\cdot a_{2i} \cdot\cdot a_{2n} \\ \cdot \\ \cdot \\ \cdot \\ a_{n1} \cdot\cdot a_{ni} \cdot\cdot a_{ni} \cdot\cdot a_{nn} \end{vmatrix} = |A| + k\cdot 0 = |A|$$

The splitting into a sum of two determinants follows from 2.21 since each term in the expansion contains just one factor from the jth column and every such element is of the form $a_{rj} + ka_{ri}$. The first determinant on the right is just $|A|$, while the second is zero by 2.32. As usual, the result for rows follows by considering A^t.

We are now in a position to give the general result of which 2.162 is a special case:

2.35 *For any $n \times n$ determinant $|A|$,*

$$\sum_{k=1}^{n} a_{jk}A_{ik} = \sum_{k=1}^{n} a_{kj}A_{ki} = 0 \qquad i \neq j$$

Proof. Consider a determinant obtained from $|A|$ by replacing the ith column by a replica of the jth column, and denote the result by $|A_0|$. Expanding $|A_0|$ according to 2.29 we have

$$|A_0| = \sum_{k=1}^{n} a_{jk}A_{ik} = 0$$

by 2.32. The second result of 2.35 follows by considering A^t.

The application of these results to the evaluation of a determinant is immediate. Our aim is to simplify the expansion by introducing as many strategically placed zeros as possible by successive applications of 2.33 and 2.34. If we can arrange that all but one element in each row and column is zero, then the evaluation will be reduced to a mere multiplication of nonzero elements, after rearrangement according to 2.31. We shall study the stages of this reduction in detail later on with reference to the *matrix* A rather than $|A|$, but we can apply it effectively here.

Example. In order to evaluate the determinant

$$\Delta = \begin{vmatrix} 1 & -1 & 1 \\ 1 & 0 & -1 \\ 0 & -2 & 1 \end{vmatrix}$$

a first step could be to subtract the first from the second row,

$$\Delta = \begin{vmatrix} 1 & -1 & 1 \\ 0 & 1 & -2 \\ 0 & -2 & 1 \end{vmatrix} = \begin{vmatrix} 1 & 0 & 0 \\ 0 & 1 & -2 \\ 0 & -2 & 1 \end{vmatrix}$$

and then, by adding the first column to the second column and subtracting it from the third, obtain zeros in the first row except in the upper left-hand corner. Again, adding twice the second row to the third,

$$\Delta = \begin{vmatrix} 1 & 0 & 0 \\ 0 & 1 & -2 \\ 0 & 0 & -3 \end{vmatrix} = \begin{vmatrix} 1 & 0 & 0 \\ 0 & 1 & 0 \\ 0 & 0 & -3 \end{vmatrix} = -3$$

and finally, adding twice the second column to the third column we obtain the desired product of single terms. Of course, we could have proceeded differently; we could indeed have stopped after the second stage and calculated the second-order determinant

$$\begin{vmatrix} 1 & -2 \\ -2 & 1 \end{vmatrix} = -3$$

While the evaluation of a determinant is relatively easy for $n = 2,3,4$, it rapidly becomes difficult, particularly if the coefficients are complicated. Using 2.31–2.35 we can now complete the

2.36 Proof of Cramer's Rule If we multiply the equation 2.181 in order by $A_{i_1}, A_{i_2}, \ldots, A_{i_n}$ and add, we have

$$\Delta x_i = \Delta_i \qquad i = 1,2, \ldots n$$

from 2.29 and 2.35, and the solutions x_i of the equations 2.181 are well determined so long as $\Delta \neq 0$.

Example. Consider the three linear equations

$$\begin{aligned} x_1 - x_2 + x_3 &= 6 \\ x_1 \qquad - x_3 &= 1 \\ - 2x_2 + x_3 &= 4 \end{aligned}$$

for which we have seen that $\Delta = -3$ in the preceding example. Applying Cramer's rule and evaluating the determinants we have

$$\Delta_1 = \begin{vmatrix} 6 & -1 & 1 \\ 1 & 0 & -1 \\ 4 & -2 & 1 \end{vmatrix} = -9, \qquad \Delta_2 = \begin{vmatrix} 1 & 6 & 1 \\ 1 & 1 & -1 \\ 0 & 4 & 1 \end{vmatrix} = 3$$

$$\Delta_3 = \begin{vmatrix} 1 & -1 & 6 \\ 1 & 0 & 1 \\ 0 & -2 & 4 \end{vmatrix} = -6$$

so that

$$x_1 = \frac{-9}{-3} = 3, \quad x_2 = \frac{3}{-3} = -1, \quad x_3 = \frac{-6}{-3} = 2$$

EXERCISES

1. Evaluate the three determinants in Exercise 1 of Section 2.1 by adding and subtracting (a) rows, (b) columns, using scalar multiplication where necessary to produce a zero element.
2. Evaluate the determinants

$$\begin{vmatrix} 1 & 1 & 2 & 1 \\ 0 & 2 & 3 & -4 \\ 3 & 2 & 1 & 0 \\ 1 & -1 & 1 & 1 \end{vmatrix}, \quad \begin{vmatrix} 1 & -1 & 3 & -1 \\ 2 & -2 & -1 & 3 \\ 1 & -1 & -4 & 4 \\ 5 & -5 & -6 & 10 \end{vmatrix}, \quad \begin{vmatrix} 1 & 1 & 1 & 5 \\ 0 & 0 & 1 & 2 \\ 3 & 4 & -3 & 2 \\ 1 & -2 & 4 & 5 \end{vmatrix}$$

3. Solve the following system of equations by Cramer's rule:

$$\begin{aligned} x \qquad + 3z + w &= 2 \\ x + 2y + 2z - w &= -2 \\ 2x + 3y + z + w &= 2 \\ x - 4y \qquad + w &= 2 \end{aligned}$$

How could you have deduced the result by inspection?

4. Prove that

$$\begin{vmatrix} a & b & c \\ a^2 & b^2 & c^2 \\ a^3 & b^3 & c^3 \end{vmatrix} = abc(a - b)(b - c)(c - a)$$

without expanding the determinant.

2.4 INTERSECTIONS OF THREE PLANES

We saw in Chapter 1 that a plane in a 3-dimensional Euclidean space is represented by a linear equation

2.41 $$\pi_1: \quad a_{11}x_1 + a_{12}x_2 + a_{13}x_3 = a_{10}$$

which passes through the origin if and only if $a_{10} = 0$. Let us consider the intersection of π_1 with the plane

2.42 $$\pi_2: \quad a_{21}x_1 + a_{22}x_2 + a_{23}x_3 = a_{20}$$

If the normal vectors $A_1 = (a_{11},a_{12},a_{13})$ and $A_2 = (a_{21},a_{22},a_{23})$ are *parallel*, then

$$a_{11}:a_{12}:a_{13} = a_{21}:a_{22}:a_{23}$$

and without loss of generality we can assume that $A_1 = A_2$, so that π_1 and π_2 are distinct if $a_{10} \neq a_{20}$. Two parallel planes have no common points.

If π_1 and π_2 are *not* parallel, they intersect in a line l. In order to find the equations of l we first locate a point $Z(z_1,z_2,z_3)$ on l so that

$$\begin{aligned} a_{11}x_1 + a_{12}x_2 + a_{13}x_3 &= a_{10} = a_{11}z_1 + a_{12}z_2 + a_{13}z_3 \\ a_{21}x_1 + a_{22}x_2 + a_{23}x_3 &= a_{20} = a_{21}z_1 + a_{22}z_2 + a_{23}z_3 \end{aligned}$$

(e.g., we may set $x_3 = 0$ and solve 2.41 and 2.42). Having thus determined Z, we may write 2.41 and 2.42 in the form

2.411 $$a_{11}(x_1 - z_1) + a_{12}(x_2 - z_2) + a_{13}(x_3 - z_3) = 0$$

2.421 $$a_{21}(x_1 - z_1) + a_{22}(x_2 - z_2) + a_{23}(x_3 - z_3) = 0$$

from which we obtain the parametric equations of l:

2.422 $$\begin{aligned} x_1 &= z_1 + \tau A_{31} \\ x_2 &= z_2 + \tau A_{32} \\ x_3 &= z_3 + \tau A_{33} \end{aligned}$$

where

$$A_{31} = \begin{vmatrix} a_{12} & a_{13} \\ a_{22} & a_{23} \end{vmatrix}, \qquad A_{32} = -\begin{vmatrix} a_{11} & a_{13} \\ a_{23} & a_{21} \end{vmatrix}, \qquad A_{33} = \begin{vmatrix} a_{11} & a_{12} \\ a_{21} & a_{22} \end{vmatrix}$$

are the direction numbers of l. Since we have assumed that π_1 and π_2 are not parallel, it follows that no two of A_{31}, A_{32}, A_{33} are zero.

Example. The two planes

$$\begin{aligned} x_1 - x_2 + 2x_3 &= 0 \\ x_1 + x_2 - x_3 &= 2 \end{aligned}$$

intersect in a line l through the point $Z(1,1,0)$, so we may write the two equations in the form

$$\begin{aligned} (x_1 - 1) - (x_2 - 1) + 2x_3 &= 0 \\ (x_1 - 1) + (x_2 - 1) - x_3 &= 0 \end{aligned}$$

Thus we have

$$x_1 = 1 - \tau, \quad x_2 = 1 + 3\tau, \quad x_3 = 2\tau$$

or in symmetric form

$$\frac{x_1 - 1}{1} = \frac{x_2 - 1}{3} = \frac{x_3}{2}$$

Consider now the intersection of π_1, π_2 and a third plane

2.43 $$\pi_3\colon \quad a_{31}x_1 + a_{32}x_2 + a_{33}x_3 = a_{30}$$

If we assume that $\Delta = |A|$ and Δ_1, Δ_2, Δ_3 are as defined in Section 2.1, then by Cramer's rule

2.44 *The coordinates of the point of intersection of the three planes π_1, π_2, π_3 are given by*

$$\Delta x_1 = \Delta_1, \quad \Delta x_2 = \Delta_2, \quad \Delta x_3 = \Delta_3$$

provided $\Delta \neq 0$.

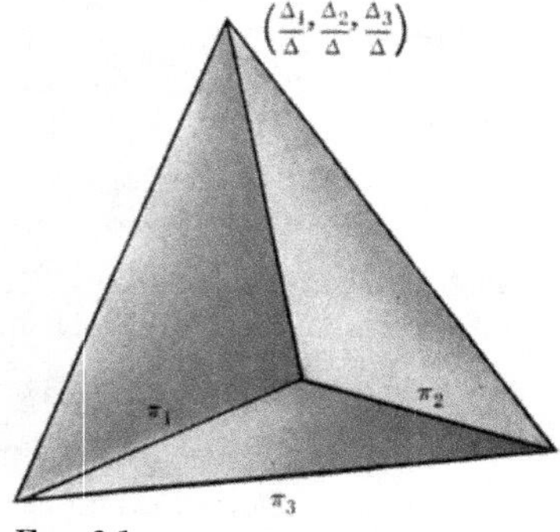

FIG. 2.1

If we assume that $\Delta = 0$ and *not all of the cofactors A_{11}, A_{21}, A_{31} are zero*, then from 2.29 and 2.35

$$\begin{aligned} a_{11}A_{11} + a_{21}A_{21} + a_{31}A_{31} &= 0 \\ a_{12}A_{11} + a_{22}A_{21} + a_{32}A_{31} &= 0 \\ a_{13}A_{11} + a_{23}A_{21} + a_{33}A_{31} &= 0 \end{aligned}$$

But these are just the scalar equations equivalent to the vector equation

$$\boldsymbol{A}_1A_{11} + \boldsymbol{A}_2A_{21} + \boldsymbol{A}_3A_{31} = 0$$

where $\boldsymbol{A}_i$ is the ith row vector of A which defines the normal to the plane π_i. We conclude that *these normal vectors are linearly dependent, and so coplanar* by 1.66. In other words, the planes π_1, π_2, π_3 are parallel to a fixed line l, perpendicular to the plane containing their normals. Put in this way we analyze the possibilities as follows.

2.45 *The three planes π_1, π_2, π_3 intersect in a line l.* In this case the three planes are linearly *dependent* and we may write

$$\pi_1 A_{11} + \pi_2 A_{21} + \pi_3 A_{31} = 0$$

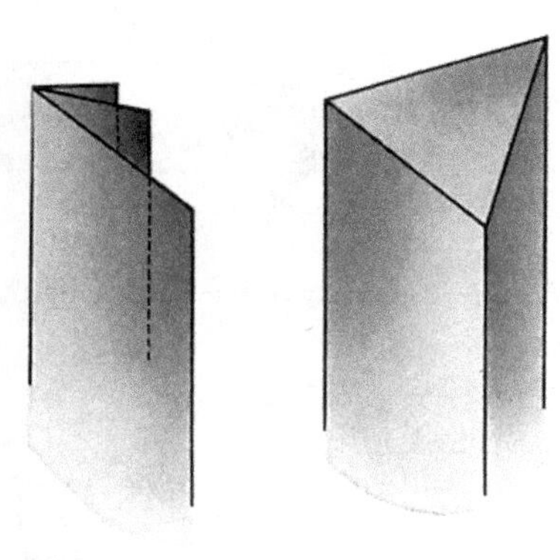

FIG. 2.2

This is equivalent to saying that in addition to the three equations written above we also have

$$a_{10}A_{11} + a_{20}A_{21} + a_{30}A_{31} = 0$$

so that the three rows of the nonsquare matrix

$$\tilde{A} = \begin{pmatrix} a_{11} & a_{12} & a_{13} & a_{10} \\ a_{21} & a_{22} & a_{23} & a_{20} \\ a_{31} & a_{32} & a_{33} & a_{30} \end{pmatrix}$$

are linearly dependent (note that we do *not* associate a determinant with a nonsquare matrix). The equations of l are given in 2.422, and the first two row vectors of $\tilde{A}$ are linearly *independent.* We say that the *row rank* of each of A and $\tilde{A}$ is 2.

2.46 *The three planes π_1, π_2, π_3 intersect in three parallel lines.* The direction numbers of these lines may be taken to be the co-factors in $|A|$, since

$$a_{11}A_{11} + a_{12}A_{12} + a_{13}A_{13} = 0$$
$$a_{21}A_{11} + a_{22}A_{12} + a_{23}A_{13} = 0$$

by 2.35. It follows that

$$A_{11}:A_{12}:A_{13} = A_{31}:A_{32}:A_{33}$$

and similarly

$$A_{21}:A_{22}:A_{23} = A_{31}:A_{32}:A_{33}$$

The row rank of A is still 2 but that of $\tilde{A}$ is now 3.

Example. The three planes (for which $\Delta = 0$),

$$\begin{aligned} x_1 - x_2 + 2x_3 &= -1 \\ x_1 + x_2 - x_3 &= 1 \\ 5x_1 + x_2 + x_3 &= k \end{aligned}$$

intersect in a line with direction numbers $(-1,3,2)$ if $k = 1$, and otherwise in three parallel lines.

2.47 *The three planes are parallel.* In this case we may assume that the row vectors of A coincide so that

$$A_1 = A_2 = A_3$$

but that $a_{10} \neq a_{20} \neq a_{30}$. The row rank of A is now 1 while that of $\tilde{A}$ is 2. We could think of this as a limiting case of 2.45.

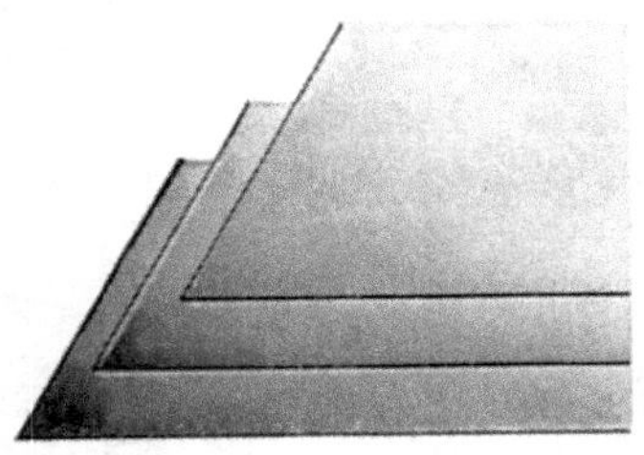

FIG. 2.3

2.48 *Two of the three planes are parallel.* We may suppose that $A_1 = A_2 \neq A_3$ so that this is a limiting case of 2.46. The row ranks of A and $\tilde{A}$ are again 2 and 3. In order to distinguish this case from 2.46, we note that here

$$A_{31} = A_{32} = A_{33} = 0$$

whereas no such set A_{k1}, A_{k2}, A_{k3} vanishes in 2.46.

There remains the case in which $\Delta = 0$ with $A_{11} = A_{21} = A_{31} = 0$, so that the last two *column* vectors of A are linearly dependent. Since the direction numbers of the lines of intersection of π_1, π_2, π_3 are again proportional, we have a special case of 2.45 (2.46) in which the line (lines) is (are) parallel to a coordinate plane, or to one of the coordinate axes.

Example. In order to illustrate this special case of 2.45 where $\Delta = 0 = A_{11} = A_{21} = A_{31}$, we take the equations

$$\begin{aligned} x_1 - x_2 + 2x_3 &= -1 \\ 2x_1 + x_2 - 2x_3 &= 1 \\ 3x_1 + 2x_2 - 4x_3 &= 2 \end{aligned}$$

in which the last three columns of $\tilde{A}$ are proportional, and the column rank of $\tilde{A}$ is 2. The direction numbers of the line of intersection are (0,2,1). If we replaced the 2 on the right side of the last equation by, say, 0, the three planes would intersect in three lines parallel to the direction (0,2,1), as in 2.46, and the column rank of $\tilde{A}$ would be 3.

EXERCISES

1. Are there any solutions of the equations

$$x_1 + x_2 - 3 = 0, \quad 2x_1 + x_2 - 4 = 0, \quad 3x_1 - 2x_2 - 1 = 0?$$

Plot the three lines on a piece of graph paper and explain the significance of your answer geometrically. Give the geometrical condition under which a common solution should exist. What would this amount to algebraically?

2. Find the parametric equations of the line of intersection of the two planes

$$\begin{aligned} 2x_1 - x_2 + x_3 &= 1 \\ x_1 + x_2 + x_3 &= 2 \end{aligned}$$

and prove that this line is parallel to the plane

$$x_1 - 2x_2 = 3$$

3. Prove that the three planes in Exercise 2 intersect in three parallel lines. Calculate the cofactors of the matrix A as in 2.46, and verify the required proportionality relations.

4. Find the equations of the lines of intersection of the three planes

$$\begin{aligned} x_1 - x_2 + x_3 &= 1 \\ x_1 - x_2 + x_3 &= 2 \\ x_1 - 2x_2 - x_3 &= 1 \end{aligned}$$

in parametric form. Can these equations be written in symmetric form?

5. Examine the nature of the intersections of the sets of planes

(i) $x_1 - x_2 = 1, \quad x_2 - x_3 = 2, \quad 2x_1 - 2x_3 = 6$

(ii) $2x_1 - x_2 + x_3 = 0, \quad x_2 - x_3 = 1, \quad x_1 + x_3 = 2$

and determine the coordinates of all common points.

2.5 HOMOGENEOUS EQUATIONS

There is still one possibility which we have not considered, namely, that the three planes π_1, π_2, π_3 *all* pass through the origin, in which case

2.51
$$a_{10} = a_{20} = a_{30} = 0$$

and the equations 2.41, 2.42, 2.43 are said to be *homogeneous*.

If these three homogeneous equations have a solution $x_1 = a_1$, $x_2 = a_2$, $x_3 = a_3$, then $x_1 = ka_1$, $x_2 = ka_2$, $x_3 = ka_3$ is also a solution for all real values of k; the three planes intersect in a line as in 2.45, and $\Delta = 0$. Conversely, our analysis of the preceding section shows that if $\Delta = 0$ and 2.51 holds, the distinction between 2.45 and 2.46 disappears whereas 2.47 and 2.48 do not apply. Moreover, if $\Delta \neq 0$, the only solution is $x_1 = x_2 = x_3 = 0$ by Cramer's rule.

In order to generalize this result we prove first the important

2.52 Theorem *The necessary and sufficient condition that the row (column) vectors of a matrix A be linearly dependent is that $|A| = 0$.*

Proof. The necessity follows immediately since the linear relation

$$A_1 a_1 + A_2 a_2 + \ldots + A_n a_n = 0$$

where $\boldsymbol{A}_i$ is the ith row vector of A and not all the a_i vanish, implies that $\Delta = 0$ by 2.34. Similarly for the columns, if $\boldsymbol{\alpha}_i$ is the ith column vector of A,

$$\boldsymbol{\alpha}_1 a_1 + \boldsymbol{\alpha}_2 a_2 + \ldots + \boldsymbol{\alpha}_n a_n = 0$$

implies that $\Delta = 0$.

Conversely, if $\Delta = 0$ and not all cofactors A_{ij} vanish, let $A_{lk} \neq 0$. Then

$$\sum_{i=1}^{n} a_{ij} A_{ik} = 0 \qquad\qquad j = 1,2, \ldots n$$

by 2.35, so that

$$\boldsymbol{A}_1 A_{1k} + \boldsymbol{A}_2 A_{2k} + \ldots + \boldsymbol{A}_n A_{nk} = 0$$

Thus the row vectors of A are linearly dependent and the row rank of A is $<n$. By an exactly similar argument, the column rank of A is $<n$.

On the other hand, if $\Delta = 0$ and all $A_{ij} = 0$ we must use induction. Let us take as our inductive assumption that the vanishing of a determinant of order $n - 1$ implies the linear dependence of its $n - 1$ rows. Certainly this is true for $n = 2, 3$. Now consider the matrix B made up of the first $n - 1$ rows of A. If the row rank of B were equal to $n - 1$, we would have a contradiction since, by assumption, every minor of order $n - 1$ vanishes and this implies that its row rank is $<n - 1$. We conclude that the row rank of B is $<n - 1$ so that of A must be $<n$. Similarly, the column rank of A is $<n$, which proves the theorem.

We can now prove that

2.53 *The necessary and sufficient condition that a system of n homogeneous linear equations in n unknowns $x_1, x_2, \ldots x_n$ should have a solution other than $x_1 = x_2 = \ldots = x_n = 0$ is that the determinant of the coefficients have the value zero.*

Proof. The necessity of the condition follows immediately since $\Delta x_i = \Delta_i = 0$ for all i, by Cramer's rule, and if some $x_i \neq 0$ we must have $\Delta = 0$.

On the other hand, $\Delta = 0$ implies that the column vectors of A are linearly dependent by 2.52, so that there exist numbers $x_1, x_2, \ldots x_n$, not all zero, such that

$$\boldsymbol{\alpha}_1 x_1 + \boldsymbol{\alpha}_2 x_2 + \ldots + \boldsymbol{\alpha}_n x_n = 0$$

or in scalar form

2.531
$$\begin{aligned} a_{11}x_1 + a_{12}x_2 + \ldots + a_{1n}x_n &= 0 \\ a_{21}x_1 + a_{22}x_2 + \ldots + a_{2n}x_n &= 0 \\ &\ldots \\ a_{n1}x_1 + a_{n2}x_2 + \ldots + a_{nn}x_n &= 0 \end{aligned}$$

but these $x_1, x_2, \ldots x_n$ provide the nontrivial solution of the homogeneous equations 2.531 which we are seeking.

The study of analytical geometry is rewarding if we do things elegantly; otherwise it leads to a morass of ugly calculation. To suggest the elegant way, we solve the following two problems.

2.54 *Find the equation of a linear subspace through n given points.* We could take $n = 2$ or $n = 3$, but since the argument holds for any n, we consider the general case. Let the n points be $Z_k(z_1^k, z_2^k, \ldots z_n^k)$ and the required equation be

2.541
$$a_1x_1 + a_2x_2 + \ldots + a_nx_n + a_{n+1} = 0$$

so that

2.542
$$a_1z_1^k + a_2z_2^k + \ldots + a_nz_n^k + a_{n+1} = 0 \qquad (k = 1,2, \ldots n)$$

We may consider 2.541 along with 2.542 as a set of $n + 1$ homogeneous equations in the $n + 1$ unknowns $a_1, a_2, \ldots a_n, a_{n+1}$ so that we must have

2.543
$$\begin{vmatrix} x_1 & x_2 \ldots x_n & 1 \\ z_1^1 & z_2^1 \ldots z_n^1 & 1 \\ z_1^2 & z_2^2 \ldots z_n^2 & 1 \\ \cdot & & \\ \cdot & & \\ \cdot & & \\ z_1^n & z_2^n \ldots z_n^n & 1 \end{vmatrix} = 0$$

Any examiner would accept this!—but by subtracting the last row from each of the others the determinant can be reduced to one of order n,

2.544
$$\begin{vmatrix} x_1 - z_1^n & x_2 - z_2^n \ldots x_n - z_n^n \\ z_1^1 - z_1^n & z_2^1 - z_2^n \ldots z_n^1 - z_n^n \\ \cdot & \\ \cdot & \\ \cdot & \\ z_1^{n-1} - z_1^n & z_2^{n-1} - z_2^n \ldots z_n^{n-1} - z_n^n \end{vmatrix} = 0$$

This result should be compared with 1.67.

2.55 *Find the equation of the plane through two intersecting lines*

2.551
$$a_{11}x_1 + a_{12}x_2 + a_{13}x_3 = 0 = a_{21}x_1 + a_{22}x_2 + a_{23}x_3$$

2.552
$$b_{11}x_1 + b_{12}x_2 + b_{13}x_3 = 0 = b_{21}x_1 + b_{22}x_2 + b_{23}x_3$$

The important thing is to find the vector $V(v_1,v_2,v_3)$ normal to each line. But direction numbers of 2.551 are

2.56
$$l_1 = \begin{vmatrix} a_{12} & a_{13} \\ a_{22} & a_{23} \end{vmatrix}, \quad l_2 = \begin{vmatrix} a_{13} & a_{11} \\ a_{23} & a_{21} \end{vmatrix}, \quad l_3 = \begin{vmatrix} a_{11} & a_{12} \\ a_{21} & a_{22} \end{vmatrix}$$

as in 2.422, and similar expressions yield direction numbers m_1, m_2, m_3 of 2.552. Thus if we assume the equation of the required plane is

$$v_1x_1 + v_2x_2 + v_3x_3 = 0$$

we must have

$$v_1l_1 + v_2l_2 + v_3l_3 = 0$$

and

$$v_1m_1 + v_2m_2 + v_3m_3 = 0$$

Eliminating, according to 2.53 we have the required equation:

$$\begin{vmatrix} x_1 & x_2 & x_3 \\ l_1 & l_2 & l_3 \\ m_1 & m_2 & m_3 \end{vmatrix} = 0 \tag{2.57}$$

EXERCISES

1. Discuss the solutions of the system of equations

$$x_1 + 2x_2 + 3x_3 = 0, \qquad 2x_1 - x_2 + x_3 = 0$$
$$x_1 + x_2 + 2x_3 = 0$$

and their geometrical significance.

2. Give the equation of the plane through the three points $(-1,2,0)$, $(1,0,3)$, $(-1,2,-2)$ in determinantal form and expand the determinant.

3. Write down the equation of the plane through the points $(1,1,0)$ and $(-1,0,2)$ parallel to the line $x_1 = x_2 = -x_3$.

4. Find the equation of the plane through the point $(1,2,3)$ and the line

$$x_1 - x_2 - x_3 = 0, \qquad 2x_1 + x_2 - x_3 = 2$$

5. Find the equation of the plane through the two intersecting lines

$$x_1 - x_2 + x_3 = 0 = x_1$$
$$2x_1 + x_2 - x_3 = 0 = x_2$$

6. Prove that the determinantal equation 2.543 of the linear subspace is equivalent to the vector equation

$$X = a_1Z_1 + a_2Z_2 + \ldots + a_nZ_n$$

with the condition that $a_1 + a_2 + \ldots + a_n = 1$. How would you interpret this vector equation as a set of parametric equations for the linear subspace in question?

3 MATRICES

3.1 MATRIX ADDITION AND MULTIPLICATION

In the preceding discussion of determinants and their applications to algebra and geometry, we have often found it convenient to speak of a square "array" or *matrix* apart from the calculations involved in evaluating its determinant. We also have seen that a matrix need not be square, but in this case a determinant is not defined.

Let us now consider matrices of this general form: matrices which have m rows and n columns, where $m \lesseqgtr n$. Such a matrix A is best thought of as a rectangular array of n column vectors $(a_{1j}, a_{2j}, \ldots a_{mj})$ or m row vectors $(a_{i1}, a_{i2}, \ldots a_{in})$, and we write

3.11 $$A = (a_{ij}) \qquad \begin{matrix} i = 1,2, \ldots m \\ j = 1,2, \ldots n \end{matrix}$$

If $B = (b_{ij})$, we shall write $A = B$ if and only if $a_{ij} = b_{ij}$ for all i, j.

As in the case of vectors, we can *add* matrices by simply adding corresponding row and column vectors, i.e., by adding *elements:*

3.12 $$A + B = (a_{ij}) + (b_{ij}) = (a_{ij} + b_{ij})$$

Also, we can define the multiplication of a matrix A by a scalar k:

3.13 $$kA = Ak = (ka_{ij})$$

In particular, for $k = 0$ we have the *zero matrix*, all of whose elements are zero.

In order to see the significance that can be attached to a matrix and to suggest how we may define the multiplication of two matrices, we take the following simple calculation from everyday life.

Example. A housewife goes to market to buy 1 lb. coffee at 75¢/lb., ½ lb. cheese at 60¢/lb., 2 lb. butter at 50¢/lb., 1 doz. oranges at 40¢/doz., and 3 loaves bread at 20¢/loaf. Her total bill is

$$1 \times 75 + \tfrac{1}{2} \times 60 + 2 \times 50 + 1 \times 40 + 3 \times 20 = \$3.05$$

which we have written as the *inner product* $\boldsymbol{A} \cdot \boldsymbol{B}$ of the two *vectors* $\boldsymbol{A}(1,\frac{1}{2},2,1,3)$ and $\boldsymbol{B}(75,60,50,40,20)$.

If we follow this suggestion by defining the *product* of two matrices A, B in terms of the inner products of their row and column vectors, then such vectors must obviously have the same number of components. Returning to the foregoing example, convention decrees that we write:

$$(1,\tfrac{1}{2},2,1,3)\begin{pmatrix}75\\60\\50\\40\\20\end{pmatrix} = (305)$$

i.e., we take the inner product of a *row* on the left with a *column* on the right.

3.14 *Definition* The *product* AB of an $m \times n$ matrix A and a $p \times q$ matrix B is an $m \times q$ matrix C if and only if $p = n$. Setting

$$A = (a_{ij}), \quad B = (b_{jk}), \quad C = (c_{ik})$$

we have

$$AB = (a_{ij})(b_{jk}) = (c_{ik}) = C$$

where

$$c_{ik} = a_{i1}b_{1k} + a_{i2}b_{2k} + \ldots + a_{in}b_{nk}$$

is the *inner product* of the ith row vector of A and the kth column vector of B.

The product AB of an $m \times n$ matrix A and a $p \times q$ matrix B is *not* defined if $p \neq n$; thus the existence of AB does not imply the existence of BA. Even if *both* AB and BA are defined, as in the case of square matrices, they are in general different, e.g.:

3.15

$$\begin{pmatrix}0 & -1\\1 & 2\end{pmatrix}\begin{pmatrix}1 & -2\\2 & 1\end{pmatrix} = \begin{pmatrix}-2 & -1\\5 & 0\end{pmatrix}$$

$$\begin{pmatrix}1 & -2\\2 & 1\end{pmatrix}\begin{pmatrix}0 & -1\\1 & 2\end{pmatrix} = \begin{pmatrix}-2 & -5\\1 & 0\end{pmatrix}$$

3.16 *Definition* If $AB = BA$, then A and B are said to *commute.*

One might imagine that definitions hedged about with so many conditions would not lead to very significant ideas. On the contrary, though we shall not be concerned very much with nonsquare matrices apart from vectors, square matrices A, B and their products will play a major role in what follows. Curiously enough, we must postpone the proof that $|AB| = |A||B|$ until we have developed more machinery.

3.2 TRANSPOSE OF A MATRIX

We have already introduced the notion of the transpose of a square matrix A obtained by interchanging the matrix rows and columns, proving in 2.27 that $|A| = |A^t|$. What is the significance of transposition for the multiplication of matrices? We prove the following important result:

3.21 *If A^t and B^t are the transposed matrices of A and B, then*

$$(AB)^t = B^tA^t$$

The following example illustrates what is going on:

$$\underset{A}{\begin{pmatrix} 0 & -1 \\ 1 & 2 \end{pmatrix}}\underset{B}{\begin{pmatrix} 1 & -2 \\ 2 & 1 \end{pmatrix}} = \underset{AB}{\begin{pmatrix} -2 & -1 \\ 4 & 0 \end{pmatrix}}, \qquad \underset{B^t}{\begin{pmatrix} 1 & 2 \\ -2 & 1 \end{pmatrix}}\underset{A^t}{\begin{pmatrix} 0 & 1 \\ -1 & 2 \end{pmatrix}} = \underset{(AB)^t}{\begin{pmatrix} -2 & 4 \\ -1 & 0 \end{pmatrix}}$$

Proof. Let us suppose that $A = (a_{ij})$, $B = (b_{kl})$ are $n \times n$ square matrices so that the product $AB = C$ is defined. Setting $C = (c_{rs})$ we have

$$c_{rs} = a_{r1}b_{1s} + a_{r2}b_{2s} + \dots + a_{rn}b_{ns} = b_{1s}a_{r1} + b_{2s}a_{r2} + \dots + b_{ns}a_{rn}$$

Since this is the element in the sth row and rth column of B^tA^t, we have proved that $B^tA^t = C^t = (AB)^t$, as desired. By successive applications, we have

3.22
$$(ABC \dots D)^t = D^t \dots C^tB^tA^t$$

EXERCISES

1. If

$$A = \begin{pmatrix} 0 & 1 & 2 \\ -1 & 0 & 1 \end{pmatrix}, \quad B = \begin{pmatrix} 1 & 0 & -1 \\ 0 & 2 & 1 \end{pmatrix}, \quad C = \begin{pmatrix} 1 & -2 & -5 \\ -2 & 2 & -1 \end{pmatrix}$$

(a) Show that $2A - B + C = 0$.

(b) Verify that $AB^t = \begin{pmatrix} -2 & 4 \\ -2 & 1 \end{pmatrix} = (BA^t)^t$.

(c) Calculate AB^tC and verify that

$$AB^tC = AB^tB - 2AB^tA$$

2. If

$$X = \begin{pmatrix} 0 & 1 & -2 \\ -1 & 0 & 1 \\ 2 & -1 & 0 \end{pmatrix}, \quad Y = \begin{pmatrix} 1 & 1 & 0 \\ -2 & 0 & 1 \\ 0 & -2 & 1 \end{pmatrix}$$

calculate X^2, XY, YX, Y^2.

3. Any matrix A is called *symmetric* if $A = A^t$. Prove that $S = \frac{1}{2}(A + A^t)$ is symmetric, and determine the matrix K such that

$$A = S + K$$

Prove that $K^t = -K$; such a matrix is called *skew-symmetric.*

4. Express each of the matrices X, Y in Exercise 2 as a sum of a symmetric matrix S and a skew-symmetric matrix K.

5. Prove that every integral power of a symmetric matrix is symmetric.

6. Prove that every *even* positive integral power of a skew-symmetric matrix is symmetric, but every *odd* positive integral power is skew-symmetric.

7. If K is an $n \times n$ skew-symmetric matrix, prove that $|K| = 0$ if n is odd.

3.3 INVERSE OF A MATRIX

If we can multiply $n \times n$ matrices, it is natural to look for an $n \times n$ matrix which, under multiplication, produces no change. Clearly, such a matrix is the *unit matrix*

3.31
$$I = \begin{pmatrix} 1 & 0 & 0 \ldots 0 \\ 0 & 1 & 0 \ldots 0 \\ \vdots & & \\ 0 & 0 & 0 \ldots 1 \end{pmatrix}$$

and $IA = AI = A$ for every $n \times n$ matrix A.

With the analogy of ordinary arithmetic in mind, it would be natural to designate the matrices

$$\begin{pmatrix} 0 & -1 \\ 1 & 2 \end{pmatrix}, \quad \begin{pmatrix} 2 & 1 \\ -1 & 0 \end{pmatrix}$$

as inverses of each other, since

$$\begin{pmatrix} 0 & -1 \\ 1 & 2 \end{pmatrix}\begin{pmatrix} 2 & 1 \\ -1 & 0 \end{pmatrix} = \begin{pmatrix} 2 & 1 \\ -1 & 0 \end{pmatrix}\begin{pmatrix} 0 & -1 \\ 1 & 2 \end{pmatrix} = \begin{pmatrix} 1 & 0 \\ 0 & 1 \end{pmatrix}$$

If we denote the *inverse* of A by A^{-1}, the question arises, does every $n \times n$ matrix A have an inverse, and is this inverse unique?

Consider the simple case where we assume that

$$A = \begin{pmatrix} 1 & -1 \\ -2 & 2 \end{pmatrix}, \qquad B = \begin{pmatrix} b_{11} & b_{12} \\ b_{21} & b_{22} \end{pmatrix}, \qquad AB = \begin{pmatrix} 1 & 0 \\ 0 & 1 \end{pmatrix}$$

so that

3.32
$$\begin{aligned} b_{11} - b_{21} &= 1, & b_{12} - b_{22} &= 0 \\ -2b_{11} + 2b_{21} &= 0, & -2b_{12} + 2b_{22} &= 1 \end{aligned}$$

Since $|A| = 0$, these equations are inconsistent and the matrix A has no inverse; A is said to be *singular*. Conversely, by Cramer's rule, the inverse A^{-1} will certainly exist if $|A| \neq 0$, in which case A is said to be *nonsingular*.

We shall give two methods of constructing the inverse A^{-1} of a nonsingular matrix A, the first along the lines of the above example. Let us suppose that $A = (a_{ij})$, $B = (b_{kl})$ are two $n \times n$ matrices such that $AB = I$ where I is the unit $n \times n$ matrix 3.31; then the following n equations determine the jth column vector $(b_{1j}, b_{2j}, \ldots b_{nj})$ of B:

3.33
$$\begin{aligned} a_{11}b_{1j} + a_{12}b_{2j} + \ldots + a_{1n}b_{nj} &= 0 \\ &\cdot \\ &\cdot \\ a_{j1}b_{1j} + a_{j2}b_{2j} + \ldots + a_{jn}b_{nj} &= 1 \\ &\cdot \\ &\cdot \\ a_{n1}b_{1j} + a_{n2}b_{2j} + \ldots + a_{nn}b_{nj} &= 0 \end{aligned}$$

In order to solve these equations 3.33 we multiply the first by A_{1i}, the second by $A_{2i}, \ldots$, and the last by A_{ni} where A_{ij} is the cofactor of a_{ij} in A; adding, every sum on the left vanishes except one:

3.34
$$(a_{1i}A_{1i} + a_{2i}A_{2i} + \ldots + a_{ni}A_{ni})b_{ij} = A_{ji}$$

But the inner product on the left is just $\Delta = |A|$ by 2.29, so that

3.35
$$b_{ij} = A_{ji}/\Delta$$

assuming that $\Delta \neq 0$. Thus, if we define the *adjoint* of A to be the matrix of cofactors

$$(A_{ij})$$

we have the inverse matrix A^{-1} of A given by

3.36
$$A^{-1} = \frac{1}{\Delta}(A_{ij})^t$$

from which we conclude that

3.37
$$(A^{-1})^t = \frac{1}{\Delta}(A_{ij}) = (A^t)^{-1}$$

Again, we encounter the problem of evaluating the determinant of a matrix. If the matrix A is large, as happens in many practical applications, the construction of A^{-1} by this method is difficult, and so we have recourse to quite a different line of thought in the following section. Before proceeding further, however, it is important to show that

3.38 $$AA^{-1} = A^{-1}A = I$$

Let us suppose that $A^{-1} = A_R^{-1}$, called the *right inverse* of A, and imagine that solving a different set of equations 3.33 would lead to a *left inverse* A_L^{-1}, where $AA_R^{-1} = A_L^{-1}A = I$. Clearly

3.39 $$A_L^{-1} = A_L^{-1}I = A_L^{-1}AA_R^{-1} = IA_R^{-1} = A_R^{-1}$$

and the *inverse* A^{-1} is uniquely defined.

EXERCISES

1. Discuss the significance of the products

$$(1,2)\begin{pmatrix} 3 \\ -1 \end{pmatrix} = (1,2)\begin{pmatrix} 5 \\ -2 \end{pmatrix} = (1,2)\begin{pmatrix} 7 \\ -3 \end{pmatrix} = (1)$$

for the existence of the inverse of a nonsquare matrix. Is there any matrix A such that $A(1,2) = (1)$?

2. Has the matrix X in Exercise 2 of the preceding section an inverse? Find the inverse of the matrix Y and verify that 3.38 holds.

3. If

$$A = \begin{pmatrix} 1 & -1 & 1 \\ 0 & 1 & -1 \\ -1 & 1 & 0 \end{pmatrix}$$

calculate A^{-1} and $(A^t)^{-1}$ and verify that $(A^{-1})^t = (A^t)^{-1}$.

3.4 REDUCTION OF A MATRIX TO CANONICAL FORM

In Chapter 2 we proved a sequence of theorems 2.31–2.34 whereby we were able to make some progress in the evaluation of a determinant. Let us try to organize the steps which were suggested in the first example of Section 2.3 into a sequence of *elementary operations*. We propose to define them with reference to a matrix A; *it is not necessary here to assume that A is square*. As a by-product of the discussion, we shall obtain a second construction for A^{-1} in this special case.

3.41 *To interchange the first two row (column) vectors of A we multiply on the left (right) by the matrix*

$$
\textbf{3.411} \qquad \left(\begin{array}{cc|ccc} 0 & 1 & 0 & & 0 \\ 1 & 0 & 0 & & 0 \\ \hline 0 & 0 & 1 & & \\ & & & 1 & \\ 0 & 0 & & & 1 \end{array}\right)
$$

If the 2×2 matrix $\begin{pmatrix} 0 & 1 \\ 1 & 0 \end{pmatrix}$ is properly placed, we may interchange any two row (column) vectors of A.

3.42 *To multiply the first row (column) vector of A by k we multiply on the left (right) by the matrix*

$$
\textbf{3.421} \qquad \left(\begin{array}{cc|ccc} k & 0 & 0 & & 0 \\ 0 & 1 & 0 & & 0 \\ \hline 0 & 0 & 1 & & \\ & & & 1 & \\ 0 & 0 & & & 1 \end{array}\right)
$$

By placing k properly we may multiply any row (column) vector of A by k.

3.43 *To add k times the first row or column vector of A to the second row or column vector we multiply on the left or right by the matrix*

$$
\textbf{3.431} \qquad \left(\begin{array}{cc|ccc} 1 & 0 & 0 & & 0 \\ k & 1 & 0 & & 0 \\ \hline 0 & 0 & 1 & & \\ & & & 1 & \\ 0 & 0 & & & 1 \end{array}\right) \quad \textit{or} \quad \left(\begin{array}{cc|ccc} 1 & k & 0 & & 0 \\ 0 & 1 & 0 & & 0 \\ \hline 0 & 0 & 1 & & \\ & & & 1 & \\ 0 & 0 & & & 1 \end{array}\right)
$$

If the 2×2 matrix $\begin{pmatrix} 1 & 0 \\ k & 1 \end{pmatrix}$ or $\begin{pmatrix} 1 & k \\ 0 & 1 \end{pmatrix}$ is properly placed, the addition of k times any given row or column vector can be made to any other row or column vector.

Matrices of the form 3.411, 3.421, 3.431 are called *elementary* and the operation accomplished by multiplying A on the left or right by such a matrix is called an *elementary operation.*

Provided not every $a_{ij} = 0$, we may, by multiplying on the left and on the right by 3.411 and 3.421, arrange that $a_{11} = 1$. When this has been accomplished, we may, by successive multiplication by matrices 3.431 on the left, arrange that $a_{i1} = 0$ for $i > 1$. After similarly arranging that $a_{22} = 1$, we may again arrange that $a_{j2} = 0$ for $j > 2$ and so on until all elements *below* the diagonal vanish. By further multiplication on the right we may arrange that every element above the diagonal vanishes also. Thus we may write

$$(P_s \ldots P_2P_1)A(Q_1Q_2 \ldots Q_t) = PAQ$$

$$= \underbrace{\begin{pmatrix} 1 & & & & & & 0 \\ & 1 & & & & & \\ & & 1 & & & & \\ & & & \cdot & & & \\ & & & & \cdot\ 1 & & \\ & & & & & 0 & \\ 0 & & & & & & 0 \end{pmatrix}}_{r_2} \Big\} r_1 \qquad (3.44)$$

where $P = P_s \ldots P_2P_1$ and $Q = Q_1Q_2 \ldots Q_t$. The matrix PAQ is now said to be in *canonical form*.

Let us consider carefully the significance of what we have done. Certainly, multiplication on the left (right) by 3.411 or 3.421 will not change the *number* of linear relations holding between the row (column) vectors of A, though the relations themselves will change. Nor will multiplication on the left (right) by 3.431 affect the number of such linear relations, though again the form of the relations will change. Thus we have proved that the *row rank and the column rank of* PAQ *are the same as the row rank* r_1 *and column rank* r_2 *of* A. But it follows from 3.44 that $r_1 = r_2 = r$, which is called the *rank* of A. We sum up these conclusions in the

3.45 Theorem *By multiplying on the left and right by suitably chosen elementary matrices* $P_s \ldots P_2P_1 = P$ *and* $Q_1Q_2 \ldots Q_t = Q$, *any* $n \times n$ *matrix* A *may be reduced to canonical form* $PAQ = I_r$, *where the number of 1's in the diagonal of* I_r *is equal to the rank* r $(\leqslant n)$ *of* A.

We illustrate this important result by the following

Example. If we suppose that

$$A = \begin{pmatrix} 2 & 4 & 0 \\ 3 & 7 & 3 \\ 0 & 2 & 6 \end{pmatrix}$$

then

$$\underset{P_3}{\begin{pmatrix} 1 & 0 & 0 \\ 0 & 1 & 0 \\ 0 & -2 & 1 \end{pmatrix}} \underset{P_2}{\begin{pmatrix} 1 & 0 & 0 \\ -3 & 1 & 0 \\ 0 & 0 & 1 \end{pmatrix}} \underset{P_1}{\begin{pmatrix} \frac{1}{2} & 0 & 0 \\ 0 & 1 & 0 \\ 0 & 0 & 1 \end{pmatrix}} A = \underset{P}{\begin{pmatrix} \frac{1}{2} & 0 & 0 \\ -\frac{3}{2} & 1 & 0 \\ 3 & -2 & 1 \end{pmatrix}} A = \begin{pmatrix} 1 & 2 & 0 \\ 0 & 1 & 3 \\ 0 & 0 & 0 \end{pmatrix}$$

so that

$$\begin{pmatrix}1 & 2 & 0\\0 & 1 & 3\\0 & 0 & 0\end{pmatrix}\underset{Q_1}{\begin{pmatrix}1 & -2 & 0\\0 & 1 & 0\\0 & 0 & 1\end{pmatrix}}\underset{Q_2}{\begin{pmatrix}1 & 0 & 0\\0 & 1 & -3\\0 & 0 & 1\end{pmatrix}} = \underset{P}{\begin{pmatrix}\frac{1}{2} & 0 & 0\\-\frac{3}{2} & 1 & 0\\3 & -2 & 1\end{pmatrix}}A\underset{Q}{\begin{pmatrix}1 & -2 & 6\\0 & 1 & -3\\0 & 0 & 1\end{pmatrix}}$$

$$= \begin{pmatrix}1 & 0 & 0\\0 & 1 & 0\\0 & 0 & 0\end{pmatrix} = I_2$$

and the rank of A is 2.

EXERCISES

1. Determine the rank of each of the matrices

$$\begin{pmatrix}1 & -1 & 2\\2 & -2 & 4\end{pmatrix}, \quad \begin{pmatrix}1 & -1 & 2\\2 & -2 & 1\end{pmatrix}$$

by applying elementary operations as described above.

2. How much of the reduction in Exercise 1 could you accomplish by operating (a) on the left *only*, (b) on the right *only*?

3. Reduce the matrix

$$\begin{pmatrix}1 & -1 & 2\\3 & 0 & 1\\-1 & 2 & 1\end{pmatrix}$$

to canonical form by applying elementary operations (a) on both sides, (b) on the left *only*, (c) on the right *only*.

4. Prove that a matrix A can be reduced to canonical form by elementary operations (a) on the left *only*, (b) on the right *only*, if A is square and also nonsingular. Are these conditions necessary as well as sufficient?

3.5 INVERSE OF A MATRIX (SECOND METHOD)

In Section 3.3 we agreed that the inverse of an $n \times n$ matrix A exists if and only if $|A| \neq 0$. Now each of the elementary matrices 3.411, 3.421, 3.431 satisfies this condition, so we may write its inverse and conclude that

3.51 $$P^{-1} = P_1^{-1}P_2^{-1} \ldots P_s^{-1}, \qquad Q^{-1} = Q_t^{-1} \ldots Q_2^{-1}Q_1^{-1}$$

Thus if A is nonsingular, $PAQ = I_r = I$ so that

3.52 $$A = P^{-1}IQ^{-1} = P^{-1}Q^{-1}$$

Since the inverse of an elementary matrix is again elementary, 3.52 expresses A as a product of elementary matrices. On the other hand, we can take the inverse of each side of 3.52 to obtain the important result

3.53 $$A^{-1} = QP$$

which yields a second and more practical method of calculating A^{-1}. However, if A is singular, $PAQ = I_r$ of rank $r < n$. Certainly $A = P^{-1}I_rQ^{-1}$, but we cannot take the inverse of I_r, so that *no expression corresponding to* **3.53** *exists if* A *is singular.*

Example. Consider the case where

$$A = \begin{pmatrix} 1 & 0 & 1 \\ 0 & -1 & 3 \\ 1 & 0 & 2 \end{pmatrix}$$

Then

$$\underset{P_2}{\begin{pmatrix} 1 & 0 & 0 \\ 0 & -1 & 0 \\ 0 & 0 & 1 \end{pmatrix}} \underset{P_1}{\begin{pmatrix} 1 & 0 & 0 \\ 0 & 1 & 0 \\ -1 & 0 & 1 \end{pmatrix}} \underset{A}{\begin{pmatrix} 1 & 0 & 1 \\ 0 & -1 & 3 \\ 1 & 0 & 2 \end{pmatrix}} \underset{Q_1}{\begin{pmatrix} 1 & 0 & -1 \\ 0 & 1 & 0 \\ 0 & 0 & 1 \end{pmatrix}} \underset{Q_2}{\begin{pmatrix} 1 & 0 & 0 \\ 0 & 1 & 3 \\ 0 & 0 & 1 \end{pmatrix}} = \begin{pmatrix} 1 & 0 & 0 \\ 0 & 1 & 0 \\ 0 & 0 & 1 \end{pmatrix}$$

so that

$$A^{-1} = \begin{pmatrix} 1 & 0 & -1 \\ 0 & 1 & 3 \\ 0 & 0 & 1 \end{pmatrix} \begin{pmatrix} 1 & 0 & 0 \\ 0 & -1 & 0 \\ -1 & 0 & 1 \end{pmatrix} = \begin{pmatrix} 2 & 0 & -1 \\ -3 & -1 & 3 \\ -1 & 0 & 1 \end{pmatrix}$$

The advantage of this method of calculating A^{-1} is just that each "elimination" is explicit and, though the method is based on Theorems 2.31 and 2.34, *no evaluations of determinants are required.* We shall have more to say on the practical computations involved after we have proved that

3.54 *If A and B are both $n \times n$ matrices, then $|AB| = |A||B|$.*

Proof. If $|B| = 0$, then the row vectors of B are linearly dependent by 2.52, and since the row vectors of AB are just linear combinations of those of B, these must also be linearly dependent and $|AB| = 0$, again by 2.52. If $|A| = 0$, then $|A^t| = 0$ by 2.27 and we can apply the same argument to $(AB)^t = B^tA^t$ to conclude that $|(AB)^t| = |AB| = 0$ as before.

If neither A nor B is singular, we base our discussion on the possibility of expressing a nonsingular matrix A as a product of elementary matrices, as in 3.52. If we could prove that

3.541 $$|CB| = |C||B|$$

for C an elementary matrix, then by breaking up and recombining, we could deduce the general result. But 3.541 follows immediately for:

(i) C of type 3.411, from 2.31 since

$$\begin{vmatrix} 0 & 1 \\ 1 & 0 \end{vmatrix} = -1$$

(ii) C of type 3.421, from 2.33 since

$$\begin{vmatrix} k & 0 \\ 0 & 1 \end{vmatrix} = k$$

(iii) C of type 3.431, from 2.34 since

$$\begin{vmatrix} 1 & 0 \\ k & 1 \end{vmatrix} = \begin{vmatrix} 1 & k \\ 0 & 1 \end{vmatrix} = 1$$

Tackling the general case, we write

$$\begin{aligned} |AB| &= |C_1C_2 \ldots C_tB| \\ &= |C_1||C_2 \ldots C_tB| = |C_1||C_2||C_3 \ldots C_tB| \\ &= |C_1C_2||C_3 \ldots C_tB| = \ldots\ldots \\ &= |C_1C_2 \ldots C_t||B| = |A||B| \end{aligned}$$

as we desired to prove.

Since $|AA^{-1}| = |I| = 1 = |A||A^{-1}|$, we have

3.55 $$|A^{-1}| = 1/|A|$$

EXERCISES

1. Find the inverse of the matrix Y in Exercise 2 of Section 3.2 by the method of this section.
2. Find the inverses of the matrices

$$A = \begin{pmatrix} 0 & 1 & 0 & 0 \\ 0 & 0 & 1 & 0 \\ 0 & 0 & 0 & 1 \\ 1 & 2 & 3 & 4 \end{pmatrix}, \quad B = \begin{pmatrix} 2 & 0 & 1 & -1 \\ 1 & -1 & 2 & 0 \\ -1 & 2 & 0 & 1 \\ 0 & 1 & -1 & 2 \end{pmatrix}$$

by the method of Section 3.3 and also by the method of this section.

3.6 THE APPROXIMATE INVERSE OF A MATRIX

In order to see the distance we have traveled since we first introduced the problem of solving the system of linear equations 2.181, let us write this system in matrix form. We have two possibilities: (i) We may consider the vector with components $x_1, x_2, \ldots x_n$ as a *column vector*, so that

3.61
$$\begin{pmatrix} a_{11} & a_{12} \ldots a_{1n} \\ a_{21} & a_{22} \ldots a_{2n} \\ \cdot & \\ \cdot & \\ \cdot & \\ a_{n1}a_{n2} \ldots a_{nn} & \end{pmatrix} \begin{pmatrix} x_1 \\ x_2 \\ \cdot \\ \cdot \\ \cdot \\ x_n \end{pmatrix} = \begin{pmatrix} a_{10} \\ a_{20} \\ \cdot \\ \cdot \\ \cdot \\ a_{n0} \end{pmatrix}$$

or (ii) we may consider the vector with components $x_1, x_2, \ldots x_n$ as a *row vector*, so that

3.62
$$(x_1, x_2, \ldots x_n) \begin{pmatrix} a_{11} & a_{21} & a_{n1} \\ a_{12} & a_{22} & a_{n2} \\ \cdot & & \\ \cdot & & \\ \cdot & & \\ a_{1n} & a_{2n} & a_{nn} \end{pmatrix} = (a_{10}, a_{20}, \ldots a_{n0})$$

It is important to observe that 3.62 is just the *transpose* of 3.61 according to 3.21, and *there is nothing to choose between the two methods of writing the equations except as convenience may dictate.*

If we denote the column vectors appearing in 3.61 by $\boldsymbol{X}$ and $\boldsymbol{\alpha}_0$, we may write the vector equation in the form

3.63
$$A\boldsymbol{X} = \boldsymbol{\alpha}_0$$

and its solution in the form

3.64
$$\boldsymbol{X} = A^{-1}\boldsymbol{\alpha}_0$$

assuming that A is nonsingular. We illustrate this *second method* of solution in the following

Example. Consider the set of equations

$$x_1 + x_2 = 1$$
$$x_1 + 1.01x_2 = 2$$

By Cramer's rule,

$$x_1 = \frac{\begin{vmatrix} 1 & 1 \\ 2 & 1.01 \end{vmatrix}}{\begin{vmatrix} 1 & 1 \\ 1 & 1.01 \end{vmatrix}} = \frac{-0.99}{0.01}, \qquad x_2 = \frac{\begin{vmatrix} 1 & 1 \\ 1 & 2 \end{vmatrix}}{\begin{vmatrix} 1 & 1 \\ 1 & 1.01 \end{vmatrix}} = \frac{1}{0.01}$$

We could write the matrix equation in the form

$$\begin{pmatrix} 1 & 1 \\ 1 & 1.01 \end{pmatrix} \begin{pmatrix} x_1 \\ x_2 \end{pmatrix} = \begin{pmatrix} 1 \\ 2 \end{pmatrix}$$

and by constructing the inverse matrix, obtain its solution in the form

$$\begin{pmatrix} x_1 \\ x_2 \end{pmatrix} = \begin{pmatrix} 101 & -100 \\ -100 & 100 \end{pmatrix} \begin{pmatrix} 1 \\ 2 \end{pmatrix} = \begin{pmatrix} -99 \\ 100 \end{pmatrix}$$

From the practical point of view this second method of determining A^{-1} is preferable for large values of n, but there is another consideration which enters into the problem. In the above example only one coefficient has been chosen to be nonintegral. If such coefficients were obtained experimentally, or were subject to an assigned "error," one might very well ask for the effect of such an error on the solution. *The answer has to do with how the inverse A^{-1} is calculated.*

In the previous section, we expressed A^{-1} as a product of elementary matrices, some on the right and some on the left. But we could have restricted ourselves to the right or the left *only* (cf. Exercises 3 and 4 of Section 3.5). Take again the matrix A of the equations of the preceding example:

$$\underbrace{\begin{pmatrix}1 & -1\\0 & 1\end{pmatrix}\begin{pmatrix}1 & 0\\0 & 100\end{pmatrix}\begin{pmatrix}1 & 0\\-1 & 1\end{pmatrix}}_{A_L^{-1}}\underbrace{\begin{pmatrix}1 & 1\\1 & 1.01\end{pmatrix}}_{A}=\begin{pmatrix}1 & 0\\0 & 1\end{pmatrix} \tag{3.65}$$

and similarly on the right. In this case $A^{-1} = A_L^{-1} = A_R^{-1}$.

If we were working approximately, we might have arrived at

$$\underbrace{\begin{pmatrix}1 & 1\\1 & 1.01\end{pmatrix}}_{A}\underbrace{\begin{pmatrix}103.01 & -99.99\\-102 & 100\end{pmatrix}}_{A_R^{-1}}=\begin{pmatrix}1.01 & 0.01\\-0.01 & 1.01\end{pmatrix} \tag{3.66}$$

which is approximately I, whereas

$$\underbrace{\begin{pmatrix}103.01 & -99.99\\-102 & 100\end{pmatrix}}_{A_R^{-1}}\underbrace{\begin{pmatrix}1 & 1\\1 & 1.01\end{pmatrix}}_{A}=\begin{pmatrix}3.02 & 2.02\\-2 & 1.01\end{pmatrix} \tag{3.67}$$

which is very different from I.

The foregoing example illustrates several things in a striking way:

(a) Matrix multiplication is *not* commutative in general.

(b) The *exact inverse* of a matrix is the same however it may be calculated (cf. 3.39).

(c) The *approximate inverse*, which is that used in actual computation, depends on *how* it is calculated, i.e., whether from the right or left. Approximate inverses calculated from both sides are particularly likely to introduce errors.*

3.7 LINEAR TRANSFORMATIONS

In 3.63, we used matrices to express a system of n linear equations as one equation. The vector $\boldsymbol{\alpha}_0$ was supposed fixed and we sought the solution vector $\mathbf{X}$. If we write the equation in the form

* Mendelsohn, N. S., "Some Elementary Properties of Ill-Conditioned Matrices and Linear Equations," *Am. Math. Monthly*, **63**, 285–295 (1956).

$$Y = AX \tag{3.71}$$

we have a *linear transformation* or *mapping* of the vectors $\boldsymbol{X}$ of $\mathcal{V}_n$ onto the vectors $\boldsymbol{Y}$; usually we shall assume that $\boldsymbol{Y}$ lies in $\mathcal{V}_n$, but it could well be a vector $\boldsymbol{X}'$ of another vector space $\mathcal{V}'_n$.

There are several important remarks concerning linear transformations which we shall make here, leaving their illustration and detailed development to the next and subsequent chapters. In the first place, the equation 3.71 could equally well be written in the form

$$Y^t = X^t A^t \tag{3.72}$$

where $\boldsymbol{Y}^t$ and $\boldsymbol{X}^t$ are *row* vectors. The two equations 3.71 and 3.72 correspond to 3.61 and 3.62.

If we think of applying the two linear transformations

$$Z = AY, \qquad Y = BX \tag{3.73}$$

in succession, we obtain as a result the linear transformation

$$Z = (AB)X \tag{3.74}$$

Actually, we have a theorem here, but the proof of 3.74 is more a matter of understanding the definition of matrix multiplication than of performing any additional mathematical operation. To clarify what is going on, we write out the steps in detail in the 2×2 case.

$$z_1 = a_{11}y_1 + a_{12}y_2, \qquad y_1 = b_{11}x_1 + b_{12}x_2$$
$$z_2 = a_{21}y_1 + a_{22}y_2, \qquad y_2 = b_{21}x_1 + b_{22}x_2$$

so that

$$\begin{aligned} z_1 &= a_{11}(b_{11}x_1 + b_{12}x_2) + a_{12}(b_{21}x_1 + b_{22}x_2) \\ &= (a_{11}b_{11} + a_{12}b_{21})x_1 + (a_{11}b_{12} + a_{12}b_{22})x_2 \\ z_2 &= a_{21}(b_{11}x_1 + b_{12}x_2) + a_{22}(b_{21}x_1 + b_{22}x_2) \\ &= (a_{21}b_{11} + a_{22}b_{21})x_1 + (a_{21}b_{12} + a_{22}b_{22})x_2 \end{aligned}$$

If we write the linear transformations 3.73 in transposed form we have:

$$Z^t = Y^t A^t, \qquad Y^t = X^t B^t \tag{3.75}$$

so that

$$Z^t = X^t(B^t A^t) = X^t(AB)^t \tag{3.76}$$

No ambiguity can arise if we drop the superscript t on the vectors $\boldsymbol{X}^t$, $\boldsymbol{Y}^t$, etc., writing 3.72 as

$$Y = XA^t \tag{3.77}$$

since *the form of writing the vectors is determined by the rule of matrix multiplication.* We shall insist, however, on calling A the *matrix* of the transformation, whether this is written in the form 3.71 or 3.77.

In Chapter 1 we introduced the notion of the basis of an n-dimensional vector space $\mathcal{V}_n$:

$$\boldsymbol{E}_1 = (1,0,\ldots 0), \quad \boldsymbol{E}_2 = (0,1,\ldots 0), \quad \ldots, \quad \boldsymbol{E}_n = (0,0,\ldots 1)$$

It is natural to ask how such vectors are transformed by the linear transformation $\boldsymbol{Y} = A\boldsymbol{X}$. We have

$$A\boldsymbol{E}_i = \begin{pmatrix} a_{11}a_{12} & \ldots & a_{1n} \\ a_{21}a_{22} & \ldots & a_{2n} \\ \cdot & & \\ a_{i1}a_{i2} & \ldots & a_{in} \\ \cdot & & \\ a_{n1}a_{n2} & \ldots & a_{nn} \end{pmatrix} \boldsymbol{E}_i = \boldsymbol{\alpha}_i$$

where $\boldsymbol{\alpha}_i$ is the ith *column* vector of A. Conversely, if we require that $\boldsymbol{E}_i$ $(i = 1,2, \ldots n)$ be transformed into $\boldsymbol{\alpha}_i$, then the matrix A is completely determined. We collect together these ideas in the following theorem.

3.78 *A linear transformation $\boldsymbol{Y} = A\boldsymbol{X}$ maps the basis vectors $\boldsymbol{E}_i$ $(i = 1,2, \ldots n)$ on the column vectors $\boldsymbol{\alpha}_i$ of A. These vectors $\boldsymbol{\alpha}_i$ are linearly independent if, and only if, A is nonsingular. Conversely, the linear transformation is completely determined when the vectors $\boldsymbol{\alpha}_i$ are given.*

It is natural to call the linear transformation 3.71 or 3.77 singular or nonsingular according to whether A is singular or nonsingular. It follows immediately that the inverse of $\boldsymbol{Y} = A\boldsymbol{X}$ is $\boldsymbol{X} = A^{-1}\boldsymbol{Y}$ and, in transposed form, that the inverse of $\boldsymbol{Y} = \boldsymbol{X}A^t$ is $\boldsymbol{X} = \boldsymbol{Y}(A^t)^{-1} = \boldsymbol{Y}(A^{-1})^t$ by 3.37.

EXERCISES

1. Write the following system of equations in the form 3.61 and solve by constructing the matrix A^{-1}:

$$\begin{aligned} x_1 - x_2 + x_3 + x_4 &= 2, & 3x_1 + 2x_2 + x_3 + x_4 &= 0 \\ 2x_1 + x_2 + x_3 + x_4 &= 1, & 4x_1 \qquad + 2x_3 + 3x_4 &= 3 \end{aligned}$$

2. If basis vectors $\boldsymbol{E}_i$ $(i = 1,2,3,4)$ are transformed by a linear transformation $\boldsymbol{Y} = A\boldsymbol{X}$ into vectors $\boldsymbol{\alpha}_1(1,2,3,4)$, $\boldsymbol{\alpha}_2(-1,1,2,0)$, $\boldsymbol{\alpha}_3(1,1,1,2)$, $\boldsymbol{\alpha}_4(1,1,1,3)$ respectively, find the vector $\boldsymbol{Y}$ into which the vector $\boldsymbol{X}(1,0,1,0)$ is transformed.

3. If the matrix A is defined as in Exercise 2, find the vector $\boldsymbol{Y}$ into which $\boldsymbol{X}(1,0,1,0)$ is transformed by the linear transformation (a) $\boldsymbol{Y} = A^2\boldsymbol{X}$, (b) $\boldsymbol{Y} = \boldsymbol{X}A$.

4 GROUPS AND LINEAR TRANSFORMATIONS

4.1 DEFINITION OF A GROUP*

If everyone in a gathering of students were asked to give his definition of mathematics, many would define *algebra* and some *geometry*; to the economist, mathematics is *arithmetic* or *statistics*, and to the engineer it is almost certainly the *calculus*. Possibly they might all agree on the definition: Mathematics is the study of *numbers* and their properties. But what are numbers but abstractions from the world around us? The *two*-ness of a pair of apples or a pair of oranges provides a starting point for a satisfactory definition of number! So perhaps we had better settle on the statement: Mathematics is the science of *abstraction*. For example, the equation of a parabola can yield the path of a projectile or the shape of a reflector on a motor headlamp. Were it not for the abstracting process,

* This section is taken from a lecture given by the author to students in Australia and published in the Year Book (1959) of the Sydney University Science Association. Permission to quote is gratefully acknowledged.

i.e., mathematization, we might not have recognized the essential identity of many apparently different phenomena in the world around us.

Let us return to the most elementary mathematics and write out the laws governing the processes of *addition* and *multiplication.*

4.11 *Addition* Take the set of all integral numbers, including zero. If a, b are integral, then

(1) $a + b = c$ is also integral
(2) $(a + b) + c = a + (b + c)$ (the associative law)
(3) $a + 0 = 0 + a = a$
(4) $a + (-a) = (-a) + a = 0$

4.12 *Multiplication* Take the set of all rational numbers. If a, b are rational, then

(1) $a \times b = c$ is also rational
(2) $(a \times b) \times c = a \times (b \times c)$ (the associative law)
(3) $a \times 1 = 1 \times a = a$
(4) $a \times a^{-1} = a^{-1} \times a = 1$

We observe also that both addition and multiplication are *commutative.*

The important thing to note is that except for the interchanges of signs $+$ and $\times$, 0 and 1, *the laws are the same in both cases.* If we abstract again, we may set up a more general system called an

4.13 *Abstract Group* $\mathcal{G}$ This has as elements $G_1, G_2, \ldots,$ with a law of combination indicated by (.), such that if G_i, G_j are elements of $\mathcal{G}$, then:

(1) $G_i.G_j = G_k$ is also an element of $\mathcal{G}$
(2) $(G_i.G_j).G_k = G_i.(G_j.G_k)$ (the associative law)
(3) $G_i.G_1 = G_1.G_i = G_i$
(4) $G_i.G_i^{-1} = G_i^{-1}.G_i = G_1$

The number of elements G_i in $\mathcal{G}$ is called the *order* g of $\mathcal{G}$, and G_1 the *identity* element of $\mathcal{G}$. We do not assume commutativity; if, however, $G_i.G_j = G_j.G_i$, then the group is called *Abelian* after the mathematician Abel who first studied such a system. Could we distinguish the two Abelian groups of addition and multiplication in some way?

Suppose we define the operation S as that of adding 1 to the number 0; then we have a 1–1 correspondence,

4.14 $$n \leftrightarrow S^n$$

and we can say that the additive group of the integers is *cyclically generated* by S. On the other hand, if we denote multiplication by the prime p by the operator S_p, then every rational number can be uniquely represented in the form

4.15 $$\ldots (S_p)^{\alpha}(S_q)^{\beta}(S_r)^{\gamma} \ldots 1$$

for suitably chosen $p, q, r, \ldots$ and $\alpha, \beta, \gamma, \ldots$. This is the fundamental theorem of arithmetic. The multiplicative group of the rational numbers is generated by an *infinite* number of independent generators S_p. The order in which these are applied is unimportant, so that the group is Abelian and the direct product of cyclical groups $\{S_p\}$, $\{S_q\}$, ...

One might imagine that these simple abstractions are very old, but they were first stated in this form by Cayley only a little more than 100 years ago. Those groups which we have considered so far have been of infinite order. Let us turn now to some familiar geometrical figures whose groups of rotations are of finite order.

4.16 ***The Groups of the Regular Solids*** Symmetry of form appealed very much to the Greeks. In Plato's cosmogony, atoms of earth are represented as cubes; of fire, as tetrahedra; of air, as octahedra; and of water, as icosahedra. The dodecahedron seems to have symbolized the universe. This mystical interweaving of ideas was characteristic of Greek philosophy in which abstractions such as "the good" and "the beautiful" were objects of constant thought and discussion. The surprising thing is that having gone so far, the Greeks did not take the next step and abstract the notion of a *group*.

If we think of the rotations of a regular tetrahedron about axes (1) through the midpoints of any opposite edges, (2) through any vertex and the centroid of the opposite face, preserving the *positions* occupied by the vertices though the latter may be interchanged, we find:

3 rotations of type (1)
8 = 4 × 2 rotations of type (2)
1 the identity (i.e., no rotation at all)
12

and we say that the group of rotations of the tetrahedron is of order 12. It is worth verifying that the group of rotations of the octahedron, or the cube, is of order 24; while that of the icosahedron, or the dodecahedron, is of order 60.

There is one feature of all the groups we have considered so far which is important, namely, their *discreteness*. If we think of the motions of a chair over the floor, we can suppose the chair moved from A to B and denote this by M_{AB}; if it is then moved from B to C we could imagine it moved directly from A to C and write

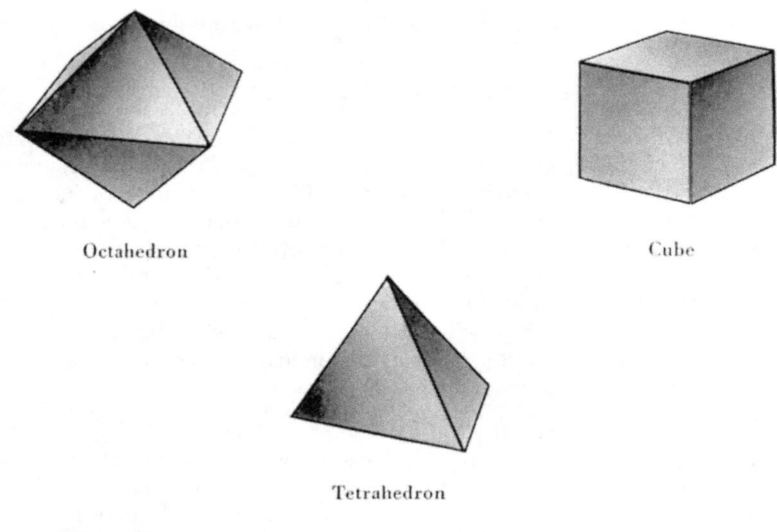

Octahedron

Cube

Tetrahedron

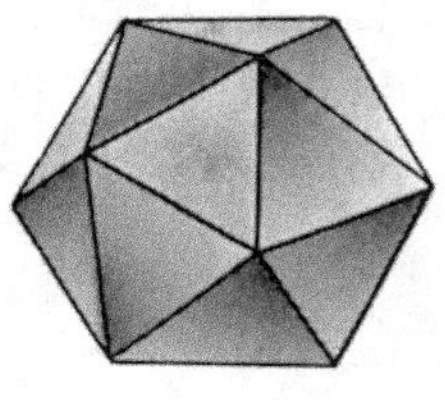

Icosahedron

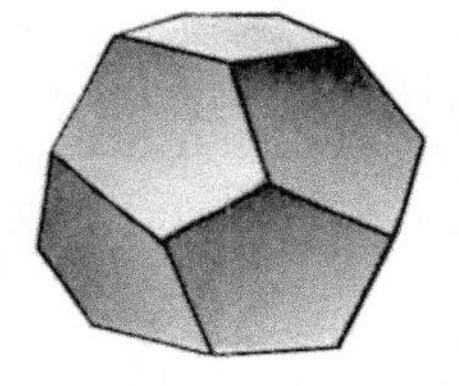

Dodecahedron

FIG. 4.1

$$M_{AB}M_{BC} = M_{AC}$$

These operations satisfy all our postulates if we let the identity operation be M_{AA}, the operation of not moving the chair at all! But here there is a difference from the groups previously considered, since we could choose B as close to A as we like. This introduces the notion of continuity, and such a group is said to be *continuous.*

4.2 THE SYMMETRIC GROUP $\mathfrak{S}_n$

In Chapter 2 we saw the significance of permutations in defining determinants and deriving their properties. In particular, in 2.24 we gave the 3! permutations on 3 symbols, and it is easy to verify that they form a group denoted by $\mathfrak{S}_3$. Since

$$(12)(23) = \begin{pmatrix}123\\132\\231\end{pmatrix} = (123), \qquad (23)(12) = \begin{pmatrix}123\\213\\312\end{pmatrix} = (132)$$

we conclude that $\mathcal{S}_3$ is *non-Abelian*, and this is true of $\mathcal{S}_n$ of order $n!$ for all $n > 2$.

If we define a *subgroup* of a given group $\mathcal{G}$ as a subset of the g elements which satisfy the conditions 4.13, it can be verified that the following list of subgroups of $\mathcal{S}_3$ is exhaustive:

4.21	I	of order 1
4.22	I, (12); I, (13); I, (23)	each of order 2
4.23	I, (123), (132)	of order 3

If $\mathcal{H}$ is any subgroup of $\mathcal{G}$ with elements $L, M, N, \ldots$, then we define the *transforms* of these elements by an element A of $\mathcal{G}$ to be ALA^{-1}, AMA^{-1}, ANA^{-1}, . . . , omitting the (.) signifying the law of combination. If $LM = N$, then

$$(ALA^{-1})(AMA^{-1}) = ALMA^{-1}$$

so that these transforms constitute a subgroup *conjugate* to $\mathcal{H}$ which we may denote $A\mathcal{H}A^{-1}$. The three subgroups 4.22 are all conjugates of one another, e.g.,

4.24 $$(13)I(13) = I, \qquad (13)(12)(13) = (23)$$

If $A\mathcal{H}A^{-1} = \mathcal{H}$ for all A in $\mathcal{G}$, $\mathcal{H}$ is said to be self-conjugate or *normal* in $\mathcal{G}$. To say that $A\mathcal{H}A^{-1} = \mathcal{H}$ means that not each element of $\mathcal{H}$ but only $\mathcal{H}$ as a set is invariant under transformation by A. For example,

4.25 $$(12)I(12) = I, \qquad (12)(123)(12) = (132)$$
$$(12)(132)(12) = (123)$$

and similarly for transformation by (13) and (23), so that the *cyclic* subgroup $\mathcal{H} = I$, (123), (132) of order 3 is normal in $\mathcal{S}_3$. No subgroup 4.22 of order 2 is normal, while the identity subgroup is always normal in any group.

In Chapter 2 we divided the $n!$ permutations on n symbols into two sets, one consisting of all the *even* permutations and the other of all the *odd* permutations. From the definition, the product of two even permutations is an even permutation. Moreover, I is even, and the inverse of an even permutation must also be even, since their product is I. Thus all the even permutations on n symbols form a subgroup of $\mathcal{S}_n$ called the *alternating group*, denoted $\mathcal{A}_n$.

On the other hand, the odd permutations do not form a subgroup of $\mathcal{S}_n$. If we multiply any even permutation of $\mathcal{A}_n$ by a single transposition, say (12), we obtain an odd permutation, e.g.,

$$\mathcal{A}_3 = I, (123), (132)$$

and

$$\mathcal{A}_3(12) = (12), (13), (23)$$

It remains to show that the number of odd permutations is exactly equal to the number of even permutations. This will follow if we can show that (i) if E_1, E_2 are even permutations and O is odd, then $E_1O \neq E_2O$ unless $E_1 = E_2$, and (ii) every odd permutation can be written in the form EO when E is even and O is a *fixed* odd permutation.

The proof of (i) is immediate, since if

$$E_1O = E_2O$$

then

$$E_1O.O^{-1} = E_2O.O^{-1}$$

so that

$$E_1 = E_2$$

assuming only that the axioms for a group are satisfied, as we have seen to be the case.

The proof of (ii) is equally easy. Let us assume O_1 to be any odd permutation and let us suppose that

$$O_1 = EO$$

as required. Since we are dealing with a group, O^{-1} exists and

$$O_1O^{-1} = EO.O^{-1} = E$$

As in (i), E is uniquely defined.

We conclude that the order of $\mathcal{A}_n$ is $\frac{1}{2}n!$ Moreover, since

$$PEP^{-1}$$

is even for all E belonging to $\mathcal{A}_n$ and *any* permutation P of $\mathcal{S}_n$, we conclude that $\mathcal{A}_n$ is normal in $\mathcal{S}_n$, generalizing 4.25. We gather together all this information in

4.26 *All even permutations on n symbols form a subgroup $\mathcal{A}_n$ of $\mathcal{S}_n$ known as the alternating group. The order of $\mathcal{A}_n$ is $\frac{1}{2}n!$ and $\mathcal{A}_n$ is normal in $\mathcal{S}_n$.*

It can be proven that $\mathcal{A}_n$ contains no normal subgroup other than I for $n \neq 4$. Such a group is said to be *simple*.

EXERCISES

1. Write out all 24 permutations of the four symbols 1,2,3,4.
2. Which ones of these are even? Verify that they form a subgroup $\mathcal{A}_4$ of $\mathcal{S}_4$.

3. Determine all subgroups of $\mathfrak{A}_4$. Which ones of these are normal (a) in $\mathfrak{A}_4$, (b) in $\mathfrak{S}_4$?

4.3 THE GROUP OF A SQUARE

Let us see if we can attach a geometrical significance to the notion of a group. To this end, consider the square in which the coordinates of the vertices $AB'A'B$ are as indicated in Figure 4.2. We say that such a geometrical figure has *symmetry*, but what precisely do we mean by this? Apart from a somewhat vague interpretation of the word, we can analyze our idea by saying that *symmetry* is characterized by the property of *invariance* under *reflection* and (or) *rotation*.

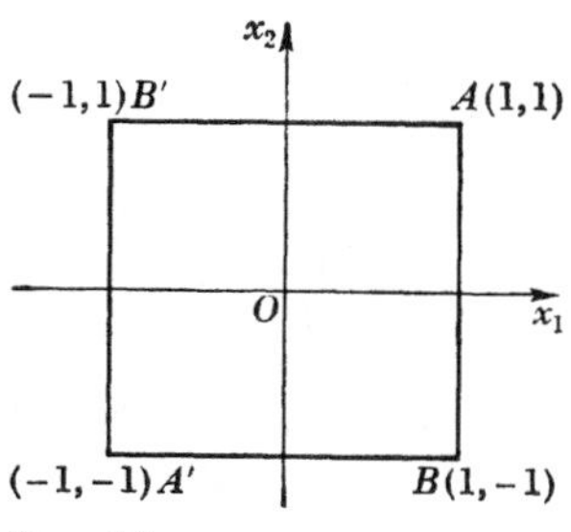

FIG. 4.2

How could we describe the operation we have called *reflection*, say in the coordinate axis Ox_2? This can be done in two ways: either as the *linear transformation*

4.311
$$\left.\begin{aligned} y_1 &= -x_1 \\ y_2 &= x_2 \end{aligned}\right\} \qquad \mathbf{Y} = \begin{pmatrix} -1 & 0 \\ 0 & 1 \end{pmatrix} \mathbf{X}$$

or relative to the square, as the *permutation*

4.312
$$(AB')(A'B)$$

Similarly, reflection in Ox_1 can be written

4.321
$$\left.\begin{aligned} y_1 &= x_1 \\ y_2 &= -x_2 \end{aligned}\right\} \qquad \mathbf{Y} = \begin{pmatrix} 1 & 0 \\ 0 & -1 \end{pmatrix} \mathbf{X}$$

or

4.322
$$(AB)(A'B')$$

The "product" of these two transformations in either order is easily seen to be

4.331
$$\left.\begin{aligned} y_1 &= -x_1 \\ y_2 &= -x_2 \end{aligned}\right\} \qquad \mathbf{Y} = \begin{pmatrix} -1 & 0 \\ 0 & -1 \end{pmatrix} \mathbf{X}$$

or

4.332
$$(AB')(A'B).(AB)(A'B') = (AA')(BB') = (AB)(A'B').(AB')(A'B)$$

and this is a *rotation* about the origin of coordinates. It can be verified that the remaining symmetries of the square are the reflections

4.341
$$\mathbf{Y} = \begin{pmatrix} 0 & -1 \\ -1 & 0 \end{pmatrix} \mathbf{X}, \qquad \mathbf{Y} = \begin{pmatrix} 0 & 1 \\ 1 & 0 \end{pmatrix} \mathbf{X}$$

or

4.342 $$(AA'), \qquad (BB')$$

and the rotations

4.351 $$Y = \begin{pmatrix} 0 & 1 \\ -1 & 0 \end{pmatrix} X, \qquad Y = \begin{pmatrix} 0 & -1 \\ 1 & 0 \end{pmatrix} X$$

or

4.352 $$(ABA'B'), \qquad (AB'A'B)$$

which, along with I,

$$\begin{pmatrix} 1 & 0 \\ 0 & 1 \end{pmatrix}$$

make the eight "operations" of the group or the square.

Not all permutations on the four symbols A, B, A', B' have geometrical significance; e.g., (AB) does not correspond to any geometrical operation valid for all points in the plane. This definition of a group of operations under which a geometrical configuration remains invariant has wide application and great importance.

EXERCISES

1. Prove that the points $A(1,0)$, $B(-\frac{1}{2},\sqrt{\frac{3}{2}})$, $C(-\frac{1}{2},-\sqrt{\frac{3}{2}})$ are the vertices of an equilateral triangle.
2. By the method of 3.78, construct the linear transformations which effect the permutations (AB) and (AC) of the vertices of the triangle in Exercise 1. Thence, construct all rotations and reflections of the group of the triangle ABC.

4.4 ROTATIONS AND REFLECTIONS

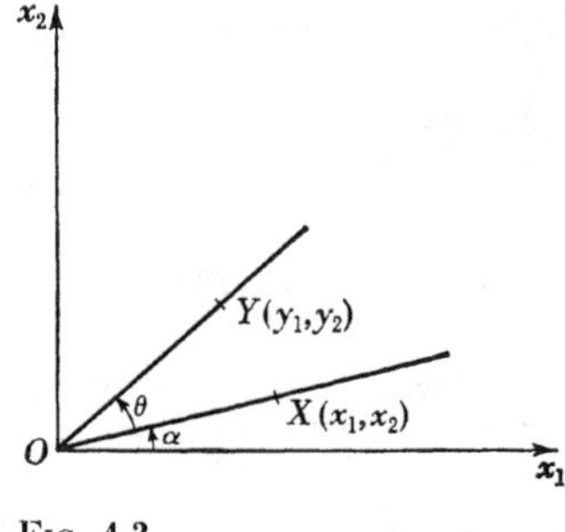

Fig. 4.3

Pursuing the line of thought of the preceding section, let us determine first the form of a *rotation* through an arbitrary angle θ about the origin. Having reference to Figure 4.3:

$$\begin{aligned} y_1 &= r\cos(\alpha+\theta) \\ &= r\cos\alpha\cos\theta - r\sin\alpha\sin\theta \\ &= x_1\cos\theta - x_2\sin\theta \\ y_2 &= r\sin(\alpha+\theta) \\ &= r\cos\alpha\sin\theta + r\sin\alpha\cos\theta \\ &= x_1\sin\theta + x_2\cos\theta \end{aligned}$$

which we can write in either one of the two ways:

$$Y = \begin{pmatrix} \cos\theta & -\sin\theta \\ \sin\theta & \cos\theta \end{pmatrix} X, \quad \text{or} \quad Y = X\begin{pmatrix} \cos\theta & \sin\theta \\ -\sin\theta & \cos\theta \end{pmatrix} \tag{4.41}$$

corresponding to 3.71 or 3.77. Observe that these formulas are independent of r and α, i.e., of the position of X. That the distance of a point from the center of rotation remains fixed is expressed in

$$y_1^2 + y_2^2 = x_1^2 + x_2^2 \tag{4.42}$$

as the *invariance* of the *quadratic* form $x_1^2 + x_2^2$ under the transformation 4.41.

If we combine two rotations, A through θ and B through ϕ, as in 3.73 and 3.74 we obtain

$$\begin{aligned} Z &= \begin{pmatrix} \cos\theta & -\sin\theta \\ \sin\theta & \cos\theta \end{pmatrix}\begin{pmatrix} \cos\phi & -\sin\phi \\ \sin\phi & \cos\phi \end{pmatrix} X \\ &= \begin{pmatrix} \cos(\theta+\phi) & -\sin(\theta+\phi) \\ \sin(\theta+\phi) & \cos(\theta+\phi) \end{pmatrix} X \end{aligned} \tag{4.43}$$

Since it would be enough if we wrote merely the *matrices* of the transformation, we are led to consider these as the elements of another group. We thus have three ways of describing the group of the square in Figure 2: (i) as a group of permutations, (ii) as a group of linear transformations, or (iii) as a group of matrices. There is an obvious one-to-one correspondence between any two of these groups, and they are said to be *isomorphic.**

There is one important point which should be emphasized. *Matrices* must be multiplied from left to right according to our definition 3.14. Consider the matrix product

$$\underset{\pi_1}{\begin{pmatrix} 0 & 1 \\ -1 & 0 \end{pmatrix}} \underset{\pi_2}{\begin{pmatrix} 0 & 1 \\ 1 & 0 \end{pmatrix}} = \underset{\pi_3}{\begin{pmatrix} 1 & 0 \\ 0 & -1 \end{pmatrix}} \tag{4.44}$$

If we write down the corresponding permutations $\pi_1 = (ABA'B')$ and $\pi_2 = (BB')$ of the preceding section, we ask the question, in what order should they be applied to yield the permutation $\pi_3 = (AB)(A'B')$? It is easy, in fact, to verify that π_2 must be applied *first* and π_1 *second* to yield π_3. If π_1 were applied first followed by π_2 we would obtain $(AB')(A'B)$ whose corresponding matrix is given in 4.311. This explains the reason

* Two groups $\mathcal{G}$, $\mathcal{G}'$ are *isomorphic* if the one-to-one correspondence between the elements $G_i \leftrightarrow G_i'$, $G_j \leftrightarrow G_j'$ extends to their combinations

$$G_i.G_j = G_k \leftrightarrow G_k' = G_i'.G_j'$$

under the appropriate laws.

for the convention of Section 2.2 that *permutations shall always be multiplied from right to left.*

Incidentally, we observe that neither the permutations nor the matrices representing them all commute, so that the group of the square is non-Abelian.

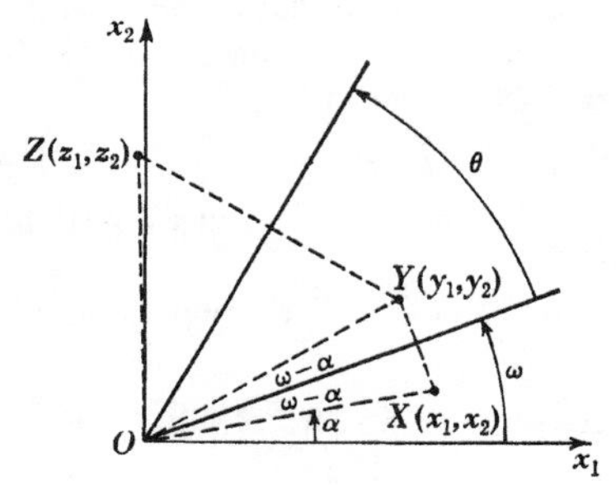

FIG. 4.4

To construct the general form of a *reflection* in an arbitrary line through the origin, we refer to Figure 4.4.

$$\begin{aligned} y_1 &= r \cos (2\omega - \alpha) \\ &= r \cos 2\omega \cos \alpha + r \sin 2\omega \sin \alpha \\ &= x_1 \cos 2\omega + x_2 \sin 2\omega \\ y_2 &= r \sin (2\omega - \alpha) \\ &= r \sin 2\omega \cos \alpha - r \cos 2\omega \sin \alpha \\ &= x_1 \sin 2\omega - x_2 \cos 2\omega \end{aligned}$$

which becomes in matrix form

4.45
$$Y = \begin{pmatrix} \cos 2\omega & \sin 2\omega \\ \sin 2\omega & -\cos 2\omega \end{pmatrix} X$$

If we reflect successively in two lines inclined to each other at an angle $\theta > 0$, the second reflection could be written:

4.46
$$Z = \begin{pmatrix} \cos (2\omega + 2\theta) & \sin (2\omega + 2\theta) \\ \sin (2\omega + 2\theta) & -\cos (2\omega + 2\theta) \end{pmatrix} Y$$

so

4.47
$$Z = \begin{pmatrix} \cos 2\theta & -\sin 2\theta \\ \sin 2\theta & \cos 2\theta \end{pmatrix} X$$

proving that:

4.48 *Successive reflection in two lines inclined at an angle* $\theta > 0$ *amounts to the same thing as rotating through* 2θ *about their common point.*

In particular, 4.311, 4.321, and 4.341 are special cases of 4.45, as is 4.331 of 4.47.

4.5 THE GROUP OF THE CUBE

All that we have said with reference to the square can easily be generalized to apply to the cube with vertices as indicated in Figure 4.5. Beginning with the reflections in the coordinate planes, we have

$$(AD')(BC')(CB')(DA'):\qquad Y = \begin{pmatrix} -1 & 0 & 0 \\ 0 & 1 & 0 \\ 0 & 0 & 1 \end{pmatrix} X$$

$$(AC')(BD')(CA')(DB'):\qquad Y = \begin{pmatrix} 1 & 0 & 0 \\ 0 & -1 & 0 \\ 0 & 0 & 1 \end{pmatrix} X$$

$$(AB')(BA')(CD')(DC'):\qquad Y = \begin{pmatrix} 1 & 0 & 0 \\ 0 & 1 & 0 \\ 0 & 0 & -1 \end{pmatrix} X$$

Note that no *one* of the four transpositions multiplied together to yield one of these permutations is a symmetry of the cube—it does not belong to the group of the cube although it does belong to the larger group $\mathcal{S}_8$ of all permutations on the eight symbols, of which the group of the cube is a subgroup.

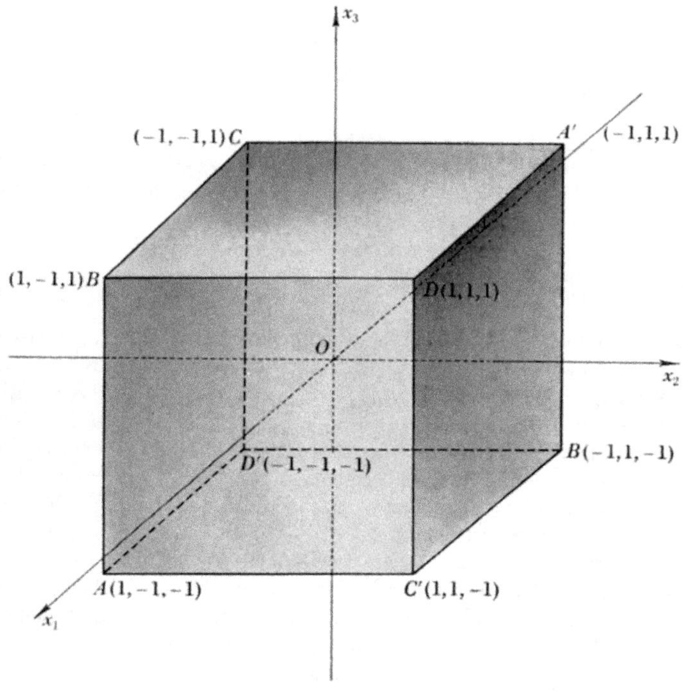

Fig. 4.5

Similarly, we may reflect in planes through pairs of opposite edges of the cube, which yields

$$(AB)(A'B'): \qquad Y = \begin{pmatrix} 0 & 1 & 0 \\ 1 & 0 & 0 \\ 0 & 0 & 1 \end{pmatrix} X$$

$$(CD)(C'D'): \qquad Y = \begin{pmatrix} 0 & -1 & 0 \\ -1 & 0 & 0 \\ 0 & 0 & 1 \end{pmatrix} X$$

Each of these reflections, and all those others that leave the cube invariant, can be constructed by applying 4.45 in the appropriate manner, i.e., by considering a reflection in space to be in a *plane* through the origin. By combining such reflections in planes we could prove the following analogue of 4.48 in space:

4.51 *Successive reflections in two planes inclined at an angle $\theta > 0$ amounts to the same thing as rotating through 2θ about their line of intersection.*

EXERCISE

1. Construct all 48 symmetries of the cube as permutations and also as linear transformations. How many of them leave invariant the regular tetrahedron *ABCD* in Figure 4.5? Prove that these form a subgroup of the group of the cube.

4.6 EULER'S FORMULA

In Figure 4.5 of the preceding section, imagine a sphere Σ drawn with center at the origin O, so as to pass through the vertices of the cube. If we project the edges of the cube from O into great circles on Σ, the resulting set of points and great circular arcs is called a *graph* and, including the faces, a *map* $\mathfrak{M}$ on Σ. What we are interested in here is not the relationship of $\mathfrak{M}$ to the cube, which is well defined in space, but the relations between the elements of $\mathfrak{M}$ on Σ. That certain of these relations remain invariant under a continuous *deformation* of the surface, suggests a new emphasis in geometry. The situation may be visualized by supposing that Σ is made of rubber and that, after the projection of the cube has been drawn on it, Σ is stretched and folded at will but not torn. The first such "topological" relation was found by Euler, but the subject was not put on a systematic basis until Poincaré's classic work of 1895.

If we denote the number of vertices of a spherical map $\mathfrak{M}$ by V, the number of edges by E, and the number of faces by F, then Euler's famous formula is that

4.61 $$V - E + F = 2$$

For 4.61 to hold for $\mathfrak{M}$, (i) the graph of $\mathfrak{M}$ must be *connected* (i.e., every vertex must be connected to every other vertex by a sequence of edges), (ii) no edge may intersect itself or any other edge except at a vertex, and (iii) every edge must be incident in exactly two faces.

The proof will be by induction. To start things off we verify the truth of 4.61 in the case of a map $\mathfrak{M}_0$ containing *one* vertex, *one* edge, and *two* faces. Such a map is obtained by taking any circle σ in Σ and a point P on σ. The circle σ divides the sphere into two faces and is itself the one edge incident in P, the one vertex of the map.

To obtain a map $\mathfrak{M}_1$ from $\mathfrak{M}_0$ we may insert a new vertex in σ, thereby *increasing* E and V each by 1. The values of V, E, F for $\mathfrak{M}_1$ would thus be $V = 2$, $E = 2$, $F = 2$.

The map $\mathfrak{M}_1$ is more symmetrical than $\mathfrak{M}_0$, in that we can add not only further vertices in the manner described above without increasing F, but also further edges, by joining pairs of vertices incident with the same face without increasing V. Such a new edge would divide the face in question so that E and F would each *increase* by 1. Each of these changes leaves 4.61 unaltered, so that any map obtainable by applying such changes in any order would satisfy 4.61 and the conditions (i) through (iii).

On the other hand, if we have given a spherical map $\mathfrak{M}$ satisfying the required conditions, we may successively remove edges which separate two faces, thereby *decreasing* E and F each by 1 until $F = 2$, always making sure that the graph remains connected. If this condition is satisfied at every stage, we may subsequently remove vertices, decreasing V and E each by 1, and reach the map $\mathfrak{M}_1$, the conditions (ii) and (iii) remaining satisfied throughout the process. It follows that:

4.62 *If a spherical map has V vertices, E edges, and F faces, and satisfies conditions* (i) *through* (iii) *above, then*

$$V - E + F = 2$$

4.7 THE REGULAR POLYHEDRA

In proving Euler's formula, we made no use of the regularity of the figure.

4.71 Definition A convex polyhedron is said to be *regular* if all its faces are regular polygons, p edges surrounding each face and q meeting in each vertex.

The notion of "convexity" is important in mathematics. Here it means that no plane containing any face of the polyhedron penetrates the interior of the polyhedron. It is natural to extend this definition of a regular polyhedron to a spherical map, and in so doing "convexity" is taken care of. The condition that p edges surround each face and q edges meet at each vertex can be written

$$pF = 2E = qV \tag{4.72}$$

since the edges are counted twice as they surround faces or meet in vertices. If we put these conditions into Euler's formula, we have

$$\frac{2E}{q} - E + \frac{2E}{p} = 2$$

or

$$\frac{1}{p} + \frac{1}{q} - \frac{1}{2} = \frac{1}{E} \tag{4.73}$$

In order to solve this Diophantine equation (i.e., an equation in more than one variable whose integral solutions are sought), we observe that $E > 0$. This limits the number of solutions to those given in the accompanying table. We recognize as corresponding to the last five solu-

4.74

	A	B	T	O	C	I	D
p	E	2	3	3	4	3	5
q	2	E	3	4	3	5	3
E	E	E	6	12	12	30	30
V	2	E	4	8	6	20	12
F	E	2	4	6	8	12	20

tions of 4.73 the *tetrahedron, octahedron, cube, icosahedron, dodecahedron* of Figure 1. The first two solutions A and B are degenerate in the sense that we have a regular map for any value of E. The first has two vertices which may be taken as north and south poles, with the E lines of longitude, equally spaced if we wish, as edges. The second is obtained by taking $V = E$ points equally spaced around an equator and counting the E intervening arcs as edges.

There is a noticeable property of the equation 4.73, namely, that from any solution we can obtain another by interchanging p and q; such *dual* solutions coincide in the case of T, for which $p = q$.

If we take the vertices A, B, C, D of the cube in Figure 5 with coordinates

$$A(1,-1,-1), \quad B(-1,1,-1), \quad C(-1,-1,1), \quad D(1,1,1)$$

it is easy to verify that $AB = AC = AD = BC = BD = 2\sqrt{2}$. The equation of the plane ABC is given by (cf. Exercise 1 of Section 1.6)

4.75
$$\begin{vmatrix} x_1 & x_2 & x_3 & 1 \\ 1 & -1 & -1 & 1 \\ -1 & 1 & -1 & 1 \\ -1 & -1 & 1 & 1 \end{vmatrix} = 0 = \begin{vmatrix} x_1+1 & x_2+1 & x_3-1 \\ 2 & 0 & -2 \\ 0 & 2 & -2 \end{vmatrix}$$

and the equations of the other three faces of the regular tetrahedron are easily found.

EXERCISES

1. Construct a cardboard model of each of the five regular polyhedra.
2. Show that the *conjugate* tetrahedron $A'B'C'D'$ has a face $A'B'C'$ parallel to ABC, by applying the symmetry of the cube:
$$\mathbf{X}' = \begin{pmatrix} -1 & 0 & 0 \\ 0 & -1 & 0 \\ 0 & 0 & -1 \end{pmatrix} \mathbf{X}$$
3. Verify that the points with coordinates
$$(0, \pm\tau, \pm 1), \quad (\pm 1, 0, \pm\tau), \quad (\pm\tau, \pm 1, 0)$$
are the vertices of a regular icosahedron if
$$\tau^2 - \tau - 1 = 0$$
4. Verify that the points with coordinates
$$(0, \pm\tau^{-1}, \pm\tau), \quad (\pm\tau, 0, \pm\tau^{-1}), \quad (\pm\tau^{-1}, \pm\tau, 0), \quad (\pm 1, \pm 1, \pm 1)$$
are the vertices of a regular dodecahedron.
5. Pick out the vertices of the five cubes which can be inscribed in the regular dodecahedron of Exercise 4.

4.8 POLYTOPES

Each regular polygon in the plane yields a regular, though degenerate, *polytope* as in the first column of the table 4.74. One might well ask what regular polyhedra exist in a space of 4 dimensions—or, more generally, in n dimensions? Curiously enough, there are *six* regular polyhedra in 4 dimensions, but for $n > 4$ there are only *three*, namely, the analogues of T, C, O.

It is easy to give the coordinates of the $2n$ vertices of the analogue of the octahedron in n dimensions:

4.81
$$(0, 0, \ldots, 0, \pm 1, 0, \ldots, 0, 0)$$

The analogue of the cube has the 2^n vertices:

$$(\pm 1, \pm 1, \ldots, \pm 1) \tag{4.82}$$

where all combinations of sign are allowed. It is best to think of the analogue of the tetrahedron or regular *simplex* in n dimensions as consisting of the $n + 1$ points

$$(0, 0, \ldots, 0, 1, 0, \ldots, 0, 0) \tag{4.83}$$

in a space of $n + 1$ dimensions. That this figure really lies in a subspace of n dimensions is shown by the fact that its vertices satisfy the linear equation

$$x_1 + x_2 + \ldots + x_n + x_{n+1} = 1 \tag{4.84}$$

It is awkward to describe the regular simplex in n dimensions by n coordinates, though this can be done also.

Two such figures make up the analogue of the octahedron in $n + 1$ dimensions, just as the corresponding two triangles

$$(1,0,0), \quad (0,1,0), \quad (0,0,1)$$

and

$$(-1,0,0), \quad (0,-1,0), \quad (0,0,-1)$$

make up the octahedron in three dimensions.

The group of symmetries of the regular simplex in n dimensions is the symmetric group $\mathfrak{S}_{n+1}$ of order $(n + 1)!$, while that of the analogue of the cube or octahedron is the *hyperoctahedral* group of order $2^n n!$.

5 VECTORS AND VECTOR SPACES

5.1 BASIS VECTORS

In Chapter 1 we introduced the notion of a vector $\mathbf{X} = \overrightarrow{OX}$ having n components $(x_1,x_2, \ldots x_n)$, and the expression

5.11 $$\mathbf{X} = x_1\mathbf{E}_1 + x_2\mathbf{E}_2 + \ldots + x_n\mathbf{E}_n$$

in terms of the basis vectors $\mathbf{E}_i$. The following result is an important consequence of 5.11:

5.12 *Any $n + 1$ vectors $\mathbf{X}, \mathbf{Y}, \ldots \mathbf{Z}$ which lie in an n-dimensional vector space $\mathcal{V}_n$ must be linearly dependent.*

Proof. If we write each vector in the form 5.11,

$$\begin{aligned} \mathbf{X} &= x_1\mathbf{E}_1 + x_2\mathbf{E}_2 + \ldots + x_n\mathbf{E}_n \\ \mathbf{Y} &= y_1\mathbf{E}_1 + y_2\mathbf{E}_2 + \ldots + y_n\mathbf{E}_n \\ &\cdot \\ &\cdot \\ &\cdot \\ \mathbf{Z} &= z_1\mathbf{E}_1 + z_2\mathbf{E}_2 + \ldots + z_n\mathbf{E}_n \end{aligned}$$

and multiply the first equation by the cofactor of $\boldsymbol{X}$ in the determinant

$$\Delta = \begin{vmatrix} \boldsymbol{X} & x_1 & x_2 & \dots & x_n \\ \boldsymbol{Y} & y_1 & y_2 & \dots & y_n \\ \cdot & & & & \\ \cdot & & & & \\ \cdot & & & & \\ \boldsymbol{Z} & z_1 & z_2 & \dots & z_n \end{vmatrix}$$

the second equation by the cofactor of $\boldsymbol{Y}$ in Δ, and finally the last equation by the cofactor of $\boldsymbol{Z}$ in Δ and add, then on the left side we have the determinant Δ. On the right, the coefficient of $\boldsymbol{E}_i$ is just Δ with $\boldsymbol{X}, \boldsymbol{Y}, \dots \boldsymbol{Z}$ replaced by $x_i, y_i, \dots z_i$, so that every such coefficient vanishes by 2.32. Thus the equation $\Delta = 0$ yields the desired linear relation between the vectors $\boldsymbol{X}, \boldsymbol{Y}, \dots \boldsymbol{Z}$ so long as not all the cofactors of $\boldsymbol{X}, \boldsymbol{Y}, \dots \boldsymbol{Z}$ vanish. But in this excluded case linear dependence follows also in virtue of 2.52

Suppose now we start with a set of n linearly independent vectors $\boldsymbol{X}_1, \boldsymbol{X}_2, \dots \boldsymbol{X}_n$ in $\mathcal{V}_n$. It follows from 5.12 that any vector $\boldsymbol{X}$ in $\mathcal{V}_n$ can be expressed

5.13 $$\boldsymbol{X} = x_1\boldsymbol{X}_1 + x_2\boldsymbol{X}_2 + \dots + x_n\boldsymbol{X}_n$$

We may describe $\boldsymbol{X}_1, \boldsymbol{X}_2, \dots \boldsymbol{X}_n$ as *basis vectors* of $\mathcal{V}_n$ and $(x_1, x_2, \dots x_n)$ as the components of $\boldsymbol{X}$ relative to this basis. Putting it otherwise, we might think of the vectors $\boldsymbol{X}_i$ as lying along n coordinate axes so that the *coordinates* of the point X would be $(x_1, x_2, \dots x_n)$. It is often convenient in analytical geometry to choose such *oblique axes*, since all the familiar intersection properties continue to hold. However, one must be careful not to interpret the Euclidean expressions for angle and distance in the usual way.

EXERCISES

1. Taking the vectors $\boldsymbol{X}_1(1,1,0,0)$, $\boldsymbol{X}_2(1,0,1,0)$, $\boldsymbol{X}_3(1,0,0,1)$, $\boldsymbol{X}_4(0,0,1,1)$ as basis, find the components of $\boldsymbol{V}(2,1,3,4)$ by solving the vector equation

$$\boldsymbol{V} = v_1\boldsymbol{X}_1 + v_2\boldsymbol{X}_2 + v_3\boldsymbol{X}_3 + v_4\boldsymbol{X}_4$$

for v_1, v_2, v_3, v_4.

2. Explain how the linear transformation

$$\boldsymbol{X} = \begin{pmatrix} 1 & 1 & 1 & 0 \\ 1 & 0 & 0 & 0 \\ 0 & 1 & 0 & 1 \\ 0 & 0 & 1 & 1 \end{pmatrix} \boldsymbol{Y}$$

relates the vectors $\boldsymbol{E}_i$ to $\boldsymbol{X}_i$. Construct the inverse transformation and therefrom derive the components of $\boldsymbol{V}$ relative to the $\boldsymbol{X}_i$, as in Exercise 1.

3. Taking $e_1(1,1)$, $e_2(1,0)$ as basis vectors defining two oblique coordinate axes Ox_1, Ox_2, make a drawing to show the positions of lines with equations

$$x_1 = 1, \quad x_2 = 1, \quad x_1 = x_2, \quad x_1 + x_2 = 1$$

Find all intersections of these lines *graphically* and verify by solving the appropriate equations.

4. Derive the linear relation connecting the four vectors $\boldsymbol{X}_1(1,1,0)$, $\boldsymbol{X}_2(1,0,1)$, $\boldsymbol{X}_3(1,0,0)$, $\boldsymbol{X}_4(0,0,1)$.

5. There may well be more than one linear relation connecting $n + 1$ vectors in $\mathcal{V}_n$. Find those connecting the vectors $\boldsymbol{X}_1(1,1,0)$, $\boldsymbol{X}_2(2,2,0)$, $\boldsymbol{X}_3(-1,0,0)$, $\boldsymbol{X}_4(2,0,0)$. What becomes of the equation $\Delta = 0$ in this case?

6. Express the number of linear relations in terms of the rank of the matrix obtained by omitting the first column of Δ.

5.2 GRAM-SCHMIDT ORTHOGONALIZATION PROCESS

Since the vectors $\boldsymbol{E}_i$ satisfy the two conditions

$$\boldsymbol{E}_i \cdot \boldsymbol{E}_j = \begin{cases} 1, & i = j \\ 0, & i \neq j \end{cases}$$

they are said to constitute a *normal, orthogonal* basis of $\mathcal{V}_n$. In general, the $\boldsymbol{X}_i$ will not satisfy such conditions and the question arises, how can we construct a normal, orthogonal basis $\boldsymbol{Y}_i$ from the $\boldsymbol{X}_i$?

Suppose we set $\boldsymbol{Y}_1 = \boldsymbol{X}_1$ and $\boldsymbol{Y}_2 = \boldsymbol{X}_2 + c\boldsymbol{Y}_1$ and require that

$$\boldsymbol{Y}_2 \cdot \boldsymbol{Y}_1 = 0 = \boldsymbol{X}_2 \cdot \boldsymbol{Y}_1 + c\boldsymbol{Y}_1 \cdot \boldsymbol{Y}_1$$

It follows that

$$c = -\frac{\boldsymbol{X}_2 \cdot \boldsymbol{Y}_1}{\boldsymbol{Y}_1 \cdot \boldsymbol{Y}_1}$$

so that

5.21
$$\boldsymbol{Y}_2 = \boldsymbol{X}_2 - \frac{\boldsymbol{X}_2 \cdot \boldsymbol{Y}_1}{\boldsymbol{Y}_1 \cdot \boldsymbol{Y}_1} \boldsymbol{Y}_1$$

Again, set $\boldsymbol{Y}_3 = \boldsymbol{X}_3 + c_2\boldsymbol{Y}_2 + c_1\boldsymbol{Y}_1$ and require that

$$\boldsymbol{Y}_3 \cdot \boldsymbol{Y}_1 = 0 = \boldsymbol{X}_3 \cdot \boldsymbol{Y}_1 + c_1\boldsymbol{Y}_1 \cdot \boldsymbol{Y}_1$$
$$\boldsymbol{Y}_3 \cdot \boldsymbol{Y}_2 = 0 = \boldsymbol{X}_3 \cdot \boldsymbol{Y}_2 + c_2\boldsymbol{Y}_2 \cdot \boldsymbol{Y}_2$$

so that

5.22
$$\boldsymbol{Y}_3 = \boldsymbol{X}_3 - \frac{\boldsymbol{X}_3 \cdot \boldsymbol{Y}_2}{\boldsymbol{Y}_2 \cdot \boldsymbol{Y}_2} \boldsymbol{Y}_2 - \frac{\boldsymbol{X}_3 \cdot \boldsymbol{Y}_1}{\boldsymbol{Y}_1 \cdot \boldsymbol{Y}_1} \boldsymbol{Y}_1$$

The procedure can be repeated indefinitely so that from any basis $\boldsymbol{X}_1, \boldsymbol{X}_2, \ldots \boldsymbol{X}_n$ we can always find an orthogonal basis $\boldsymbol{Y}_1, \boldsymbol{Y}_2, \ldots \boldsymbol{Y}_n$. It should be emphasized, however, that such an *orthogonal* basis is by no means unique.

Example. If $X_1(1,1,0)$, $X_2(1,0,1)$, $X_3(1,0,0)$ be the given basis, choose $Y_1 = X_1$,

$$Y_2 = X_2 - \tfrac{1}{2}Y_1 = (\tfrac{1}{2},-\tfrac{1}{2},1)$$

$$Y_3 = X_3 - \tfrac{1}{3}Y_2 - \tfrac{1}{2}Y_1 = (\tfrac{1}{3},-\tfrac{1}{3},-\tfrac{1}{3})$$

and these Y_i's are pairwise orthogonal, as desired.

To pass from an orthogonal basis to a *normal* orthogonal basis, it is necessary only to multiply each vector Y_i by the scalar $1/|Y_i|$. Of course, we could have normalized at each successive stage so that each of the denominators in 5.21 and 5.22 would have been 1.

EXERCISES

1. Instead of setting $Y_1 = X_1$ in the preceding example, set $Y_1 = X_3$ and complete the construction of an orthonormal basis in the usual way.
2. Set $Y_1 = X_3$, $Y_2 = X_1 - X_3$ and find Y_3 so that Y_1, Y_2, Y_3 is an orthonormal basis in the example.
3. Construct vectors Y_3, Y_4 so that along with $Y_1 = X_1$, $Y_2 = X_4$ the Y_i's constitute an *orthogonal* basis of the space of Exercise 1 of Section 5.1.
4. Using the Gram-Schmidt orthogonalization process, find an orthonormal basis Y_i for the space defined by the vectors

$$X_1(1,0,1), \quad X_2(2,-1,1), \quad X_3(-1,-1,1)$$

5.3 THE VECTOR PRODUCT $U \times V$

In the special case $n = 3$ of the preceding section,

$$Y_3 \cdot Y_1 = Y_3 \cdot X_1 = 0$$

$$Y_3 \cdot Y_2 = Y_3(X_2 + cY_1) = Y_3 \cdot X_2 = 0$$

If we set $X_1 = U$, $X_2 = V$, $Y_3 = W$, then the components of W must satisfy the scalar equations

$$W \cdot U = w_1u_1 + w_2u_2 + w_3u_3 = 0$$

$$W \cdot V = w_1v_1 + w_2v_2 + w_3v_3 = 0$$

so that (cf. 2.56)

5.31
$$w_1 : w_2 : w_3 = \begin{vmatrix} u_2 & u_3 \\ v_2 & v_3 \end{vmatrix} : \begin{vmatrix} u_3 & u_1 \\ v_3 & v_1 \end{vmatrix} : \begin{vmatrix} u_1 & u_2 \\ v_1 & v_2 \end{vmatrix}$$

Though the magnitude of $W \neq 0$ is not determined by these proportionalities, its direction is determined up to a factor of ± 1. It is customary to fix the positive direction, as in the case of the coordinate axes in Chapter 1, to be that of a "right-handed screw," under which U is rotated into V. This amounts to the following

5.32 *Definition* The vector $W = U \times V$ has components

$$w_1 = \begin{vmatrix} u_2 & u_3 \\ v_2 & v_3 \end{vmatrix}, \qquad w_2 = \begin{vmatrix} u_3 & u_1 \\ v_3 & v_1 \end{vmatrix}, \qquad w_3 = \begin{vmatrix} u_1 & u_2 \\ v_1 & v_2 \end{vmatrix}$$

It follows immediately that $(V \times U) = -(U \times V)$.

5.33 *$U \times V = 0$ if and only if the rank of the matrix*

$$\begin{pmatrix} u_1 & u_2 & u_3 \\ v_1 & v_2 & v_3 \end{pmatrix} \textit{ is } 0 \textit{ or } 1.$$

Proof. Certainly, $U \times V = 0$ if either $U = 0$ or $V = 0$ or $U = kV$ when $r = 1$; or if $U = V = 0$, then $r = 0$. Conversely, $U \times V = 0$ implies that one of these conditions must be satisfied.

In 3-space it is often convenient to set $i = E_1 = (1,0,0)$, $j = E_2 = (0,1,0)$, $k = E_3 = (0,0,1)$ so that, symbolically,

$$\textbf{5.34} \qquad U \times V = \begin{vmatrix} i & j & k \\ u_1 & u_2 & u_3 \\ v_1 & v_2 & v_3 \end{vmatrix}$$

For any W,

$$\textbf{5.35} \qquad (U \times V) \cdot W = \begin{vmatrix} u_1 & u_2 & u_3 \\ v_1 & v_2 & v_3 \\ w_1 & w_2 & w_3 \end{vmatrix} = W \cdot (U \times V)$$

and by permuting U, V, W cyclically,

$$(U \times V) \cdot W = (V \times W) \cdot U = (W \times U) \cdot V$$

$$\begin{aligned} \{(U + U') \times V\} \cdot W &= (U + U') \cdot (V \times W) \\ &= U \cdot (V \times W) + U' \cdot (V \times W) \\ &= (U \times V) \cdot W + (U' \times V) \cdot W \\ &= \{(U \times V) + (U' \times V)\} \cdot W \end{aligned}$$

Since this relation is true for all vectors W, we can choose $W = i, j, k$ in turn and, from the equal scalars on left and right, conclude that

$$\textbf{5.36} \qquad (U + U') \times V = (U \times V) + (U' \times V)$$

Thus *vector multiplication is distributive.* It is to be noted that in general

$$(U \times V) \times W \neq U \times (V \times W)$$

so that vector multiplication is *not* associative.

Since the square of the *magnitude* of $U \times V$ is given by the Lagrange identity

$$(u_2v_3 - u_3v_2)^2 + (u_3v_1 - u_1v_3)^2 + (u_1v_2 - u_2v_1)^2$$
$$= (u_1^2 + u_2^2 + u_3^2)(v_1^2 + v_2^2 + v_3^2) - (u_1v_1 + u_2v_2 + u_3v_3)^2$$

we conclude that

5.371 $$|U \times V|^2 = |U|^2|V|^2 - (U \cdot V)^2 = |U|^2|V|^2(1 - \cos^2 \theta)$$

so that

5.372 $$|U \times V| = |U||V| \sin \theta$$

where θ is the angle between U and V.

While the vector product is defined only in 3-space, we may arrive at more general ideas by observing that the Lagrange identity may be written in the form

5.373 $$|U \times V|^2 = \begin{vmatrix} U \cdot U & U \cdot V \\ V \cdot U & V \cdot V \end{vmatrix}$$

Since 5.372 leads to the interpretation of $|U \times V|$ as the *area* of the parallelogram formed by the vectors U and V, it is tempting to confine our attention to this plane π and suppose that

$$U = X(x_1,x_2), \qquad V = Y(y_1,y_2)$$

relative to some orthonormal basis in π. With such an assumption,

$$\begin{vmatrix} U \cdot U & U \cdot V \\ V \cdot U & V \cdot V \end{vmatrix} = \begin{vmatrix} X \cdot X & X \cdot Y \\ Y \cdot X & Y \cdot Y \end{vmatrix}$$

5.38 $$= \begin{vmatrix} x_1^2 + x_2^2 & x_1y_1 + x_2y_2 \\ y_1x_1 + y_2x_2 & y_1^2 + y_2^2 \end{vmatrix} = \begin{vmatrix} x_1 & x_2 \\ y_1 & y_2 \end{vmatrix} \begin{vmatrix} x_1 & y_1 \\ x_2 & y_2 \end{vmatrix}$$

$$= \begin{vmatrix} x_1 & x_2 \\ y_1 & y_2 \end{vmatrix}^2$$

by 2.27. We conclude from 5.372 and 5.373 that

5.39 *The area of the parallelogram defined by $U = X$ and $V = Y$ in π is given by*

$$|U \times V| = \begin{vmatrix} x_1 & x_2 \\ y_1 & y_2 \end{vmatrix} \quad \textit{taken positive.}$$

This remarkable simplification, which arises through consideration of the problem in a subspace of the proper dimension, *does* generalize, as we shall see shortly.

EXERCISES

1. Find the area of the parallelogram whose vertices are $(-1,1)$, $(0,0)$, $(1,3)$, $(2,2)$.
2. Find the components of $U \times V$ when $U = (1,1,0)$, $V = (1,0,1)$. Find also the area of the parallelogram defined by U and V.
3. Determine the fourth vertex of a parallelogram of which the first three vertices are

$$A(1,0,1), \quad B(-1,1,1), \quad C(2,-1,2)$$

   How many such vertices are there? What is the area of the parallelogram in each case? What is the area of the triangle ABC?
4. Find the vector $W = U \times V$ when $U = (1,-1,1)$ and $V = (1,1,-1)$, and calculate the volume of the parallelepiped defined by U, V, and W. What is the volume of the tetrahedron $OUVW$?
5. Derive the more general form of the Lagrange identity,

$$(U \times V)\cdot(U' \times V') = \begin{vmatrix} U\cdot U' & U\cdot V' \\ V\cdot U' & V\cdot V' \end{vmatrix}$$

6. Prove the Jacobi identity

$$(U \times V) \times W + (V \times W) \times U + (W \times U) \times V = 0$$

5.4 DISTANCE BETWEEN TWO SKEW LINES

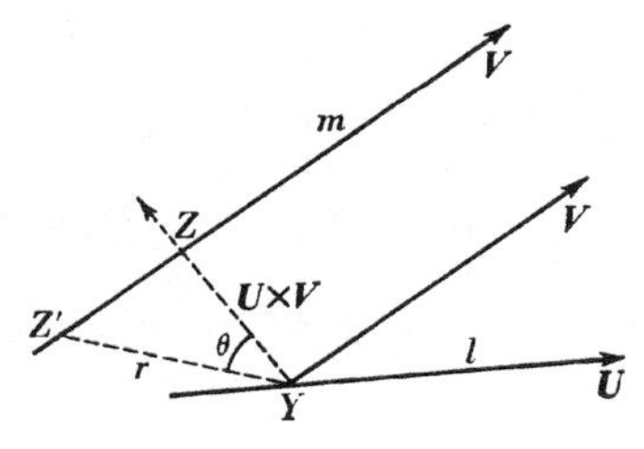

FIG. 5.1

From the geometrical point of view, the vector product $U \times V$ is just the normal vector to the plane determined by U and V and, as such, it has arisen on numerous previous occasions. By assigning a definite direction to this normal according to the definition 5.32, we have introduced a refinement which is chiefly useful in mechanics and in the study of electrical phenomena. Nevertheless, it is worth utilizing these ideas to determine the shortest distance between two skew, nonparallel lines in 3-space.

Let us take the equations of l and m in parametric form to be

$$\textit{5.41} \qquad l: \begin{aligned} x_1 &= y_1 + su_1, \\ x_2 &= y_2 + su_2, \\ x_3 &= y_3 + su_3, \end{aligned} \qquad m: \begin{aligned} x_1 &= z_1 + tv_1 \\ x_2 &= z_2 + tv_2 \\ x_3 &= z_3 + tv_3 \end{aligned}$$

and let $r = \overrightarrow{YZ}$, so that

5.42
$$r \cdot (U \times V) = \begin{vmatrix} z_1 - y_1 & z_2 - y_2 & z_3 - y_3 \\ u_1 & u_2 & u_3 \\ v_1 & v_2 & v_3 \end{vmatrix}$$

By adding multiples of the second (third) row to the first, it follows from 5.41 that the choice of Y (Z) on l (m) does not affect the value of $r \cdot (U \times V)$. Thus if we take Y to be the foot of the common perpendicular and Z to be a variable point, the expression 5.42 set equal to zero yields the equation of a *plane* through l, parallel to m. Hence, for Z on m the required perpendicular distance δ between l and m is given by

5.43
$$\delta = \frac{r \cdot (U \times V)}{|U \times V|}$$

according to 1.68. An alternative approach to the same result is indicated by writing

5.44
$$\delta = |r| \cos \theta = |r| \frac{r \cdot (U \times V)}{|r||U \times V|} = \frac{r \cdot (U \times V)}{|U \times V|}$$

where $Z = Z'$ is any point on m, as in Figure 5.1.

EXERCISES

1. Find the equations of a line drawn from the origin O to intersect each of the lines

$$l: \quad x_1 - x_2 + x_3 - 1 = 0 = 2x_1 - x_3 - 2$$
$$m: \quad x_1 - 2x_2 - 1 = 0 = x_1 - x_2 - x_3 + 5$$

(Hint: Find a plane through each of l, m which contains O.)

2. Obtain the equations of the lines l, m in Exercise 1 in parametric form and also in symmetric form.

3. Find the shortest distance between the lines l and m in Exercise 1.

5.5 n-DIMENSIONAL VOLUME

At the end of Section 5.3 we saw that the area of a parallelogram defined by two vectors X, Y in $\mathcal{V}_2$ is given by the determinant

5.51
$$\begin{vmatrix} x_1 & x_2 \\ y_1 & y_2 \end{vmatrix}$$

In order to generalize this result we shall reconsider it from a slightly different point of view.

If $\boldsymbol{W}$ is a vector coplanar with $\boldsymbol{X}$, $\boldsymbol{Y}$ and such that

$$|\boldsymbol{W}| = |\boldsymbol{Y}| \cos \phi = |\boldsymbol{Y}| \sin \left(\frac{\pi}{2} - \phi\right)$$

with

5.521 $$\boldsymbol{W} \cdot \boldsymbol{X} = 0$$

then

5.522 $$\boldsymbol{W} \cdot \boldsymbol{Y} = |\boldsymbol{W}| \cdot |\boldsymbol{Y}| \cos \phi = |\boldsymbol{W}|^2$$

which we may write in detail as follows:

$$w_1 x_1 + w_2 x_2 = 0$$

$$w_1 y_1 + w_2 y_2 = |\boldsymbol{W}|^2$$

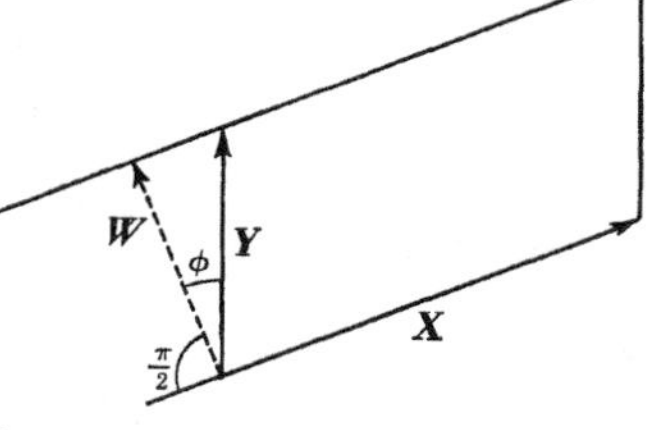

Fig. 5.2

Solving these equations, we have by Cramer's rule

5.53 $$\frac{w_1}{\begin{vmatrix} 0 & x_2 \\ |\boldsymbol{W}|^2 & y_2 \end{vmatrix}} = \frac{w_2}{\begin{vmatrix} x_1 & 0 \\ y_1 & |\boldsymbol{W}|^2 \end{vmatrix}} = \frac{1}{\begin{vmatrix} x_1 & x_2 \\ y_1 & y_2 \end{vmatrix}} = \frac{1}{\begin{vmatrix} x_1 & x_2 \\ w_1 & w_2 \end{vmatrix}}$$

the last fraction being obtained by multiplying numerator and denominator of the first by w_1 and the second by w_2, adding, and dividing out the factor $|\boldsymbol{W}|^2$.

It is important to realize that *area is invariant under rotation.* To see this, we suppose the axes transformed as in 4.41 so that

$$\begin{vmatrix} x_1 & x_2 \\ y_1 & y_2 \end{vmatrix} = \begin{vmatrix} x_1 & x_2 \\ y_1 & y_2 \end{vmatrix} \begin{vmatrix} \cos\theta & \sin\theta \\ -\sin\theta & \cos\theta \end{vmatrix} = \begin{vmatrix} x_1' & x_2' \\ y_1' & y_2' \end{vmatrix}$$

and also

$$\begin{vmatrix} x_1 & x_2 \\ y_1 & y_2 \end{vmatrix} = \begin{vmatrix} x_1 & x_2 \\ w_1 & w_2 \end{vmatrix} \begin{vmatrix} \cos\theta & \sin\theta \\ -\sin\theta & \cos\theta \end{vmatrix} = \begin{vmatrix} x_1' & x_2' \\ 0 & w \end{vmatrix} = w x_1'$$

where θ is determined by the condition that

$$w_1 \cos\theta - w_2 \sin\theta = 0$$

and $w = w_1 \sin\theta + w_2 \cos\theta$. Since $w x_1'$ is the usual expression for the area of the parallelogram in question, we may use induction to obtain the required generalization of 5.51, if we assume that *volume and generalized volume are similarly invariant under rotation.* Further properties and applications of the important "orthogonal" transformation involved here will be found in Chapter 9.

As before, let $\boldsymbol{W}$ be a vector cospatial with $\boldsymbol{X}$, $\boldsymbol{Y}$, $\boldsymbol{Z}$ and such that $|\boldsymbol{W}| = |\boldsymbol{Z}| \cos \phi$,

5.541 $$\boldsymbol{W} \cdot \boldsymbol{X} = 0$$

5.542 $$\boldsymbol{W}\cdot\boldsymbol{Y} = 0$$

and

5.543 $$\boldsymbol{W}\cdot\boldsymbol{Z} = |\boldsymbol{W}|\cdot|\boldsymbol{Z}| \cos\phi = |\boldsymbol{W}|^2$$

Writing out these equations in full:

$$\begin{aligned} w_1x_1 + w_2x_2 + w_3x_3 &= 0 \\ w_1y_1 + w_2y_2 + w_3y_3 &= 0 \\ w_1z_1 + w_2z_2 + w_3z_3 &= |\boldsymbol{W}|^2 \end{aligned}$$

we solve to obtain

$$\frac{w_1}{\begin{vmatrix} 0 & x_2 & x_3 \\ 0 & y_2 & y_3 \\ |\boldsymbol{W}|^2 & z_2 & z_3 \end{vmatrix}} = \frac{w_2}{\begin{vmatrix} x_1 & 0 & x_3 \\ y_1 & 0 & y_3 \\ z_1 & |\boldsymbol{W}|^2 & z_3 \end{vmatrix}} = \frac{w_3}{\begin{vmatrix} x_1 & x_2 & 0 \\ y_1 & y_2 & 0 \\ z_1 & z_2 & |\boldsymbol{W}|^2 \end{vmatrix}}$$

5.55 $$= \frac{1}{\begin{vmatrix} x_1 & x_2 & x_3 \\ x_1 & y_2 & y_3 \\ z_1 & z_2 & z_3 \end{vmatrix}} = \frac{1}{\begin{vmatrix} x_1 & x_2 & x_3 \\ y_1 & y_2 & y_3 \\ w_1 & w_2 & w_3 \end{vmatrix}}$$

Again the denominator of the last fraction is independent of $\boldsymbol{Z}$, and after a suitable transformation we may suppose that $\boldsymbol{W}' = (0,0,w)$, so that this denominator becomes

$$w\begin{vmatrix} x_1' & x_2' \\ y_1' & y_2' \end{vmatrix}$$

Since this is, by definition, the volume of the parallelepiped defined by the vectors $\boldsymbol{X}$, $\boldsymbol{Y}$, $\boldsymbol{Z}$, we conclude that this volume is given by the determinant

5.56 $$\begin{vmatrix} x_1 & x_2 & x_3 \\ y_1 & y_2 & y_3 \\ z_1 & z_2 & z_3 \end{vmatrix}$$

in every case.

Having established the basis of our induction and the definition of a generalized volume in $\mathcal{V}_n$ in terms of that in $\mathcal{V}_{n-1}$, we have proved the following

5.57 Theorem *The generalized volume of the generalized parallelepiped defined by n linearly independent vectors* $\boldsymbol{X}_i$ $(i = 1,2, \ldots n)$ *in* $\mathcal{V}_n$ *with components* $(x_{i1}, x_{i2}, \ldots x_{in})$ *is given by the determinant*

$$\begin{vmatrix} x_{11} & x_{12} & \cdots & x_{1n} \\ x_{21} & x_{22} & \cdots & x_{2n} \\ \cdot & & & \\ \cdot & & & \\ \cdot & & & \\ x_{n1} & x_{n2} & \cdots & x_{nn} \end{vmatrix}$$

It follows from its determinantal expression that the generalized volume is zero if the n vectors are linearly dependent. On the other hand, if the n linearly independent vectors lie in $\mathcal{V}_m$ $(m > n)$, the matrix of components is no longer square and the determinant is no longer defined. As in 5.38, however, we observe that

5.58
$$\begin{vmatrix} \mathbf{X}_1\cdot\mathbf{X}_1 & \mathbf{X}_1\cdot\mathbf{X}_2 & \cdots & \mathbf{X}_1\cdot\mathbf{X}_n \\ \mathbf{X}_2\cdot\mathbf{X}_1 & \mathbf{X}_2\cdot\mathbf{X}_2 & \cdots & \mathbf{X}_2\cdot\mathbf{X}_n \\ \vdots & & & \\ \mathbf{X}_n\cdot\mathbf{X}_1 & \mathbf{X}_n\cdot\mathbf{X}_2 & \cdots & \mathbf{X}_n\cdot\mathbf{X}_n \end{vmatrix}$$

$$= \begin{vmatrix} x_{11} & x_{12} & \cdots & x_{1n} \\ x_{21} & x_{22} & \cdots & x_{2n} \\ \vdots & & & \\ x_{n1} & x_{n2} & \cdots & x_{nn} \end{vmatrix} \begin{vmatrix} x_{11} & x_{21} & \cdots & x_{n1} \\ x_{12} & x_{22} & \cdots & x_{n2} \\ \vdots & & & \\ x_{1n} & x_{2n} & \cdots & x_{nn} \end{vmatrix}$$

and the determinant of inner products 5.58 is defined in $\mathcal{V}_m$ for every $m \geqslant n$. Thus:

5.59 *The square of the generalized volume of the generalized parallelepiped defined by n linearly independent vectors $\mathbf{X}_i$ $(i = 1,2, \ldots n)$ in $\mathcal{V}_m$ $(m \geqslant n)$ is given by the determinant*

$$\begin{vmatrix} \mathbf{X}_1\cdot\mathbf{X}_1 & \mathbf{X}_1\cdot\mathbf{X}_2 & \cdots & \mathbf{X}_1\cdot\mathbf{X}_n \\ \mathbf{X}_2\cdot\mathbf{X}_1 & \mathbf{X}_2\cdot\mathbf{X}_2 & \cdots & \mathbf{X}_2\cdot\mathbf{X}_n \\ \vdots & & & \\ \mathbf{X}_n\cdot\mathbf{X}_1 & \mathbf{X}_n\cdot\mathbf{X}_2 & \cdots & \mathbf{X}_n\cdot\mathbf{X}_n \end{vmatrix}$$

As before, this determinantal expression vanishes if the n vectors are linearly dependent.

EXERCISES

1. Find the volume of the parallelepiped defined by the three vectors $\mathbf{X}(0,0,1)$, $\mathbf{Y}(0,1,2)$, $\mathbf{Z}(1,2,3)$.

2. Make a drawing of the figure in Exercise 1; find the area A of the parallelogram in the plane $x_1 = 0$, and calculate the required volume by means of the formula $V = A|\mathbf{Z}| \cos \theta_1$.

3. Find the volume of the parallelepiped defined by the three vectors $\mathbf{X}(0,0,1,1)$, $\mathbf{Y}(0,1,2,2)$, $\mathbf{Z}(1,2,3,3)$.

4. Find the area of the face defined by X and Y in Exercise 3, and using the method of Exercise 2 calculate the required volume.
5. What is the volume of the tetrahedron $OXYZ$ in Exercises 1 and 3?
6. Could you set up an induction which would yield the volume of an n-dimensional simplex in terms of the volume of the generalized parallelepiped in 5.59?

5.6 SUBSPACES OF $\mathcal{V}_n$

Though a geometrical entity such as the *volume* of a figure may not change its value, its description relative to the space in which it is embedded may change, as we have seen. We have this phenomenon arising in a simpler form in the case of a line, which is defined by one linear equation in 2-space, by two linear equations in 3-space, . . . by $n - 1$ linear equations in n-space. Thus the *dimension* of the space in which a geometrical figure is embedded is important in describing it analytically.

Let $\mathbf{X}_1, \mathbf{X}_2, \ldots \mathbf{X}_m$ be any m linearly independent vectors of $\mathcal{V}_n$ so that $m \leqslant n$ by 5.12. If $0 < m < n$, we say that these m vectors define a subspace $\mathcal{V}_m$ of $\mathcal{V}_n$ made up of all vectors

5.61
$$\mathbf{X} = \sum_{i=1}^{m} a_i \mathbf{X}_i$$

Any vector $\mathbf{Y}$ which cannot be written in this form does not belong to $\mathcal{V}_m$. Since there are just n linearly independent vectors in $\mathcal{V}_n$, we may choose as basis

5.62
$$\mathbf{X}_1, \mathbf{X}_2, \ldots \mathbf{X}_m, \quad \mathbf{Y}_1, \mathbf{Y}_2, \ldots \mathbf{Y}_{n-m}$$

and we may suppose that the Gram-Schmidt orthogonalization process has been applied so that they are all pairwise orthogonal. Since every $\mathbf{X}_i \cdot \mathbf{Y}_j = 0$, it follows that

5.63
$$\left(\sum_{i=1}^{m} a_i \mathbf{X}_i\right) \cdot \left(\sum_{1}^{n-m} b_j \mathbf{Y}_j\right) = 0$$

and *every vector in* $\mathcal{V}_m$ *is orthogonal to every vector in the subspace* $\mathcal{V}_{n-m}$ *defined by the vectors* $\mathbf{Y}_1, \mathbf{Y}_2, \ldots \mathbf{Y}_{n-m}$.

The two subspaces $\mathcal{V}_m$ and $\mathcal{V}_{n-m}$ are said to be *orthogonal complements* of each other relative to $\mathcal{V}_n$.

Example. For $n = 2$, any two distinct vectors are linearly independent, and if orthogonal they are complementary relative to $\mathcal{V}_2$. But two orthogonal vectors are not complementary in a three-dimensional vector space $\mathcal{V}_3$, whereas a plane and its normal vector define complementary subspaces in $\mathcal{V}_3$. In $\mathcal{V}_4$ the orthogonal complement of a $\mathcal{V}_1$ is a $\mathcal{V}_3$ and that of a $\mathcal{V}_2$ is another $\mathcal{V}_2$.

Just as we broke down $\mathcal{V}_n$ into orthogonal complementary subspaces $\mathcal{V}_m$ and $\mathcal{V}_{n-m}$, so we could break down $\mathcal{V}_m$. The particular case in which $\mathcal{V}_{2n}$ is broken down into n pairwise orthogonal planes, or $\mathcal{V}_{2n+1}$ into n pairwise orthogonal planes and a line orthogonal to each plane, is of special interest. If we think of a rotation in each such plane about the complementary subspace we have, taking them all together, the most general rotation in $\mathcal{V}_{2n}$ or $\mathcal{V}_{2n+1}$, as we shall see later on.

Two subspaces $\mathcal{V}_r$ and $\mathcal{V}_s$ of $\mathcal{V}_n$ may intersect. We define the subspace made up of all vectors common to $\mathcal{V}_r$ and $\mathcal{V}_s$ as the *intersection* $\mathcal{V}_r \cap \mathcal{V}_s$, and the subspace made up of all vectors linearly dependent on vectors of $\mathcal{V}_r$ and $\mathcal{V}_s$ as the *union* $\mathcal{V}_r \cup \mathcal{V}_s$. If we denote the dimensions of $\mathcal{V}_r \cap \mathcal{V}_s$ and $\mathcal{V}_r \cup \mathcal{V}_s$ by $d(\mathcal{V}_r \cap \mathcal{V}_s)$ and $d(\mathcal{V}_r \cup \mathcal{V}_s)$, then

5.64
$$d(\mathcal{V}_r \cup \mathcal{V}_s) + d(\mathcal{V}_r \cap \mathcal{V}_s) = r + s$$

Proof. If $\mathcal{V}_r$ and $\mathcal{V}_s$ have no vectors in common, then $d(\mathcal{V}_r \cap \mathcal{V}_s) = 0$ and the number of linearly independent vectors in $\mathcal{V}_r \cup \mathcal{V}_s$ is just $r + s$, as claimed. If, however, $d(\mathcal{V}_r \cap \mathcal{V}_s) > 0$, then we may suppose that the $d(\mathcal{V}_r \cap \mathcal{V}_s) = d$ linearly independent vectors $\mathbf{Z}_1, \mathbf{Z}_2, \ldots \mathbf{Z}_d$ of $\mathcal{V}_r \cap \mathcal{V}_s$ form part of the basis of each of $\mathcal{V}_r$ and $\mathcal{V}_s$:

5.65
$$\overbrace{\mathbf{X}_1, \mathbf{X}_2, \ldots \mathbf{X}_{n-d},\ \mathbf{Z}_1, \mathbf{Z}_2, \ldots \mathbf{Z}_d,}^{\mathcal{V}_r}\ \mathbf{Y}_1, \mathbf{Y}_2, \ldots \mathbf{Y}_{s-d} \qquad \underbrace{\mathbf{Z}_1, \mathbf{Z}_2, \ldots \mathbf{Z}_d,\ \mathbf{Y}_1, \mathbf{Y}_2, \ldots \mathbf{Y}_{s-d}}_{\mathcal{V}_s}$$

so that in the enumeration every vector $\mathbf{Z}_i$ is counted twice on each side of 5.64, proving the result.

If we replace $\mathcal{V}_r$ and $\mathcal{V}_s$ by their orthogonal complements $\mathcal{V}_{n-r}$ and $\mathcal{V}_{n-s}$, 5.64 becomes

5.66
$$[n - d(\mathcal{V}_r \cup \mathcal{V}_s)] + [n - d(\mathcal{V}_r \cap \mathcal{V}_s)] = (n - r) + (n - s)$$

as was to be expected.

5.7 EQUATIONS OF A SUBSPACE

From the point of view of analytical geometry, all the subspaces of $\mathcal{V}_n$ pass through the origin and so are defined by one or more *homogeneous* linear equations. For example, $\mathcal{V}_m$ is defined by the $n - m$ linear equations

5.71
$$\begin{aligned} a_{11}x_1 + a_{12}x_2 + \ldots + a_{1n}x_n &= 0 \\ a_{21}x_1 + a_{22}x_2 + \ldots + a_{2n}x_n &= 0 \\ &\vdots \\ a_{n-m,1}x_1 + a_{n-m,2}x_2 + \ldots + a_{n-m,n}x_n &= 0 \end{aligned}$$

To see that this is so, it is only necessary to adjoin the m further equations

$$\begin{aligned} x_{n-m+1} &= t_1 \\ x_{n-m+2} &= t_2 \\ &\vdots \\ x_n &= t_m \end{aligned}$$

and solve by Cramer's rule to obtain $(x_1, x_2, \ldots x_n)$ expressed in terms of the *parameters* $t_1, t_2, \ldots t_m$. Since these expressions are linear, just m solutions are linearly independent and we can associate them with the values of the parameters,

$$1,0,0, \ldots 0; \quad 0,1,0, \ldots 0; \quad \ldots; \quad 0,0,0, \ldots 1$$

If we call the corresponding vectors $\mathbf{X}_1, \mathbf{X}_2, \ldots \mathbf{X}_m$, we have recovered $\mathcal{V}_m$ as the *solution space* of the set of linear equations 5.71.

Example. Let us suppose that $n = 3$, and we wish to find a basis of the solution space of the equation

5.72 $$2x_1 - 3x_2 + x_3 = 0$$

To this we adjoin the equations

$$x_2 = t_1, \qquad x_3 = t_2$$

so that

5.73 $$x_1 = \tfrac{3}{2}t_1 - \tfrac{1}{2}t_2$$

for all t_1, t_2. As above, the basis vectors of $\mathcal{V}_2$ could be taken to be

5.74 $$\mathbf{X}_1 = (\tfrac{3}{2},1,0), \qquad \mathbf{X}_2 = (-\tfrac{1}{2},0,1)$$

Certainly these are linearly independent and any solution of 5.72 defines a vector

$$\mathbf{X} = t_1\mathbf{X}_1 + t_2\mathbf{X}_2$$

Conversely, every such vector yields a solution of 5.72.

Thus the dimension of the solution space is 2, i.e., 5.72 represents a plane $\mathcal{V}_2$ through the origin containing the vectors $\mathbf{X}_1, \mathbf{X}_2$ and every vector linearly dependent on them. These vectors form a basis of $\mathcal{V}_2$, from which an orthonormal basis could be constructed by the Gram-Schmidt orthogonalization process.

If we wish to study the intersection $\mathcal{V}_2 \cap \mathcal{V}_2'$, where $\mathcal{V}_2'$ is defined by the equation

5.75 $$x_1 + x_2 - x_3 = 0$$

we should adjoin *one* further equation, $x_3 = t_1$, and solve 5.72 and 5.75 to obtain

5.76 $$x_1 = \tfrac{2}{5}t_1, \quad x_2 = \tfrac{3}{5}t_1, \quad x_3 = t_1$$

Thus the intersection space has dimension 1 with basis vector $Z = (2,3,5)$. To bring this vector into evidence as in 5.65, we observe that it arises by setting $t_1 = 3$, $t_2 = 5$ in 5.73. Clearly, any vector of $\mathcal{V}_2$ may be written in the form

5.77 $$x_1 = -t_1' + t_2', \quad x_2 = t_1', \quad x_3 = t_2'$$

from which we obtain Z again by setting $t_1' = 3$, $t_2' = 5$. As a basis for $\mathcal{V}_2 \cup \mathcal{V}_2' = \mathcal{V}_3$, we could choose

5.78 $$X_1 = (\tfrac{3}{2},1,0), \quad Z = (2,3,5), \quad Y_1 = (1,0,1)$$

since Y_1 does not lie in $\mathcal{V}_2$. We note in passing that the zero vector lies in every subspace, since the defining equations are all homogeneous.

5.8 ORTHOGONAL PROJECTION

If we take $E_1 = (1,0)$, $E_2 = (0,1)$ as basis vectors in $\mathcal{V}_2$, then the coordinate axes are two mutually orthogonal subspaces and any vector

$$X = x_1E_1 + x_2E_2$$

is said to have *orthogonal projections* x_1E_1 and x_2E_2 on these axes. More generally, if X_1 and Y_1 are orthogonal, then any vector X may similarly be written

$$X = a_1X_1 + b_1Y_1$$

and X has orthogonal projections a_1X_1 and b_1Y_1 on X_1 and Y_1.

If we take the basis 5.62 of $\mathcal{V}_n$, then any vector X of $\mathcal{V}_n$ may be written in the form

$$X = \sum_{i=1}^{m} a_iX_i + \sum_{j=1}^{n-m} b_jY_j$$

and we say that X has orthogonal projections

$$\sum_{i=1}^{m} a_iX_i \quad \text{on} \quad \mathcal{V}_m$$

and

$$\sum_{j=1}^{n-m} b_jY_j \quad \text{on} \quad \mathcal{V}_{n-m}$$

Since any subspace of $\mathcal{V}_n$ is defined in terms of its basis vectors, we may similarly speak of its orthogonal projections on $\mathcal{V}_n$ and $\mathcal{V}_{n-m}$, but we shall not pursue the matter further.

EXERCISES

1. Find an orthogonal basis for vectors which lie in the plane $x_1 - x_2 + x_3 = 0$. Find also the orthogonal projection of the vector $(1,0,1)$ on this plane and on the normal to the plane.

2. Find parametric equations and orthogonal basis vectors for the solution space of the system of equations

$$\begin{aligned} x_1 - x_2 - x_3 + 2x_4 &= 0 \\ 2x_1 - x_2 - 2x_3 + x_4 &= 0 \end{aligned}$$

3. Prove that the vector (1,1,0,0) does *not* lie in the solution space of Exercise 2, and find its orthogonal projection (a) on that space, (b) on each of the primes represented by the two equations which define the space.

6 CONICS AND QUADRICS

6.1 CIRCLES AND SPHERES

So far we have concerned ourselves with linear properties of algebra and geometry, but Pythagoras' theorem suggests that we should study *quadratic* properties also, since the relation between the sides of a triangle is not between their lengths but between the *squares* of these lengths. In fact, this theorem yields the analytic definition of a *circle* of radius r and center (a_1,a_2) to be

6.11
$$(x_1 - a_1)^2 + (x_2 - a_2)^2 = r^2$$

and that of a *sphere* with center (a_1,a_2,a_3) to be

6.12
$$(x_1 - a_1)^2 + (x_2 - a_2)^2 + (x_3 - a_3)^2 = r^2$$

Expanding these expressions we observe that (i) *the coefficients of the square terms x_i^2 are all equal,* and (ii) *no cross terms x_ix_j appear.* Conversely, if these conditions are satisfied the squares may be completed and the equation takes the form 6.11 in the plane and 6.12 in space.

It follows immediately that circles and spheres, like lines and planes, have the property that any two define a *linear pencil*. Consider, for example, the two circles

$$C_a: \quad (x_1 - a_1)^2 + (x_2 - a_2)^2 - r_a^2 = 0$$
$$C_b: \quad (x_1 - b_1)^2 + (x_2 - b_2)^2 - r_b^2 = 0$$

Any linear combination

6.13 $$C_a + \rho C_b = 0$$

satisfies the conditions (i) and (ii) above, so that 6.13 represents a circle. Moreover, any point satisfying $C_a = 0 = C_b$ also satisfies $C_a + \rho C_b = 0$ so that if the two circles $C_a = 0$ and $C_b = 0$ intersect in P, Q then every circle of the pencil passes through P, Q. Two circles intersect in *two* points, which may coincide, or not at all since elimination leads to a quadratic equation which has two real roots or two complex roots. In the latter case no two circles of the pencil intersect. Such a linear pencil of circles is said to be *coaxal*, and we illustrate the three possibilities in Figure 6.1.

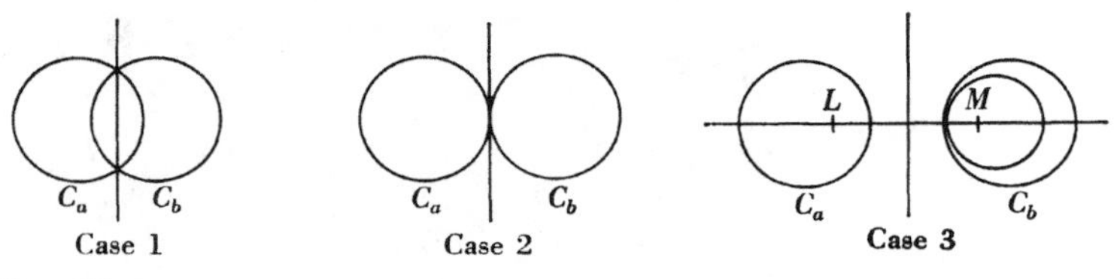

FIG. 6.1

If $\rho = -1$, 6.13 reduces to a linear equation which represents the line PQ in Case 1 and Case 2 and a line not meeting any circle of the coaxal pencil in Case 3; this line is called the *radical axis* of the system.

In Case 1 the diameter of a circle belonging to the linear pencil cannot be less than $|PQ|$, whereas no such restriction holds in Case 2 or Case 3. In the latter case we may let the radius tend to zero and define two *limit points* L, M which are equidistant from the radical axis. The following example illustrates Case 3.

Example. Take the two circles in Case 3 to be

$$C_a: \quad (x_1 - 2)^2 + x_2^2 = 1$$
$$C_b: \quad (x_1 + 1)^2 + x_2^2 = 1$$

The equation $C_a + \rho C_b = 0$ when simplified becomes

$$(1 + \rho)x_1^2 + (-4 + 2\rho)x_1 + (1 + \rho)x_2^2 + 3 = 0$$

and if we require the radius of such a circle to be zero, then $\rho^2 - 7\rho + 1 = 0$ so that $\rho = \frac{1}{2}(7 \pm 3\sqrt{5})$. It follows that the x_1 coordinates of the *limit points* L, M (circles of zero radius) are $-\frac{1}{2}(-1 + \sqrt{5}) = -0.62$ and $\frac{1}{2}(1 + \sqrt{5}) = 1.62$, with $x_2 = 0$ in each case. The equation of the radical axis is $x_1 = \frac{1}{2}$.

All this generalizes to spheres in space and indeed to hyperspheres in any number of dimensions. Any three circles, spheres, or hyperspheres of a coaxal system are linearly *dependent* in the sense we have so often used the term.

If we suppose that $a_1 = a_2 = 0$ in 6.11, we have the equation of a circle with center the origin,

6.14 $$x_1^2 + x_2^2 = r^2$$

Alternatively, we could have applied the linear transformation

6.15 $$\begin{aligned} y_1 &= x_1 - a_1 \\ y_2 &= x_2 - a_2 \end{aligned}$$

called a "parallel translation" or simply a *translation* to achieve the simplification of the equation. We can think of this as moving the circle 6.11 while keeping the axes fixed, or as a change of axes. We shall study such transformations in Chapter 9, but we note in passing that the equation 6.14 remains *invariant* under any rotation or reflection 4.41 or 4.45.

More generally, we may write the equation of a hypersphere in $\mathcal{V}_n$, with center the origin, in the form

6.16 $$X^tIX \equiv x_1^2 + x_2^2 + \ldots + x_n^2 = r^2$$

Any linear transformation which leaves 6.16 invariant is said to be *orthogonal*, and we shall see in Chapter 9 that such a transformation may be thought of as a succession of rotations and reflections, when these are suitably defined. We prove here the important theorem

6.17 *The necessary and sufficient condition that a linear transformation* $Y = AX$ *should be orthogonal is that* $A^{-1} = A^t$.

Proof. If we assume that

$$Y^tY = X^tA^tAX = X^tIX$$

then $A^tA = I$ and $A^{-1} = A^t$ as required. Conversely, this condition is sufficient as well as necessary for the invariance of the *quadratic form* $X^tIX = X^tX$.

EXERCISES

1. Prove that the centers of the circles of the coaxal system 6.13 are collinear and find the equation of the line in question.

2. Prove that the radical axis of a coaxal system of nonintersecting circles is the right bisector of LM, where L, M are the limit points of the system.

3. Prove that every circle with center on the radical axis and passing through the limit points cuts every circle of the system orthogonally.

4. Find the point of intersection P of the radical axes of the three circles C_a, C_b of the above example along with C_c: $x_1^2 + (x_2 - 2)^2 - 1 = 0$, taken two by two. The point P is called the *radical center* of the system of circles. Prove that the circle with center P, passing through the limit points, cuts any circle

$$C_a + \rho C_b + \sigma C_c = 0$$

orthogonally.

5. Write the equation of a circle through the three points (0,1), (1,0), (−2,3) as a determinant.

6. Prove that the equation of a circle through the three points (0,1), (1,0), $(\frac{1}{2},\frac{1}{2})$, when written in determinantal form, reduces to $x_1 + x_2 = 1$. Explain.

7. Write the equation of a sphere through the four points (0,1,1), (1,0,1), (1,1,0), (1,1,1) in determinantal form and find its radius.

8. If

$$1 + a^2 - b^2 - c^2 = \alpha, \qquad 1 - a^2 + b^2 - c^2 = \beta$$
$$1 - a^2 - b^2 + c^2 = \gamma, \qquad 1 + a^2 + b^2 + c^2 = \delta$$

verify that the matrix

$$A = \delta^{-1}\begin{pmatrix} \alpha & 2(ab + c) & 2(ac - b) \\ 2(ab - c) & \beta & 2(bc + a) \\ 2(ac + b) & 2(bc - a) & \gamma \end{pmatrix}$$

due to Rodrigues and Euler, is *orthogonal* for all real values of a, b, c.

9. Show by actual substitution that the linear transformation $\mathbf{Y} = A\mathbf{X}$, with $n = 3$, obtained by setting $a = b = 1$, $c = 0$ in Exercise 8, leaves $\mathbf{X}^t\mathbf{X}$ invariant.

6.2 CONICS IN CARTESIAN COORDINATES

Besides the circle in the plane, there are two other central *conics* called the *ellipse*,

6.21
$$\frac{x_1^2}{a^2} + \frac{x_2^2}{b^2} = 1$$

and the *hyperbola*,

6.22
$$\frac{x_1^2}{a^2} - \frac{x_2^2}{b^2} = 1$$

The matrix equations for these curves are easily seen to be

6.23
$$\mathbf{X}^t\begin{pmatrix}\frac{1}{a^2} & 0\\ 0 & \frac{1}{a^2}\end{pmatrix}\mathbf{X} = 1, \qquad \mathbf{X}^t\begin{pmatrix}\frac{1}{a^2} & 0\\ 0 & \frac{-1}{b^2}\end{pmatrix}\mathbf{X} = 1$$

which correspond to 6.16 for a circle.

It is not our purpose here to investigate the properties of these *central conics* in great detail. Suffice it to say that they are *symmetrical* with respect to each coordinate axis and so with respect to the origin which is called the *center* of the conic. The ellipse 6.21 cuts each axis in two pairs of *vertices* $(\pm a,0)$, $(0,\pm b)$, the lines joining these pairs of vertices being called the *principle axes* of the conic. The length of the *semimajor* axis is a and that of the *semiminor* axis is b, if $a > b$.

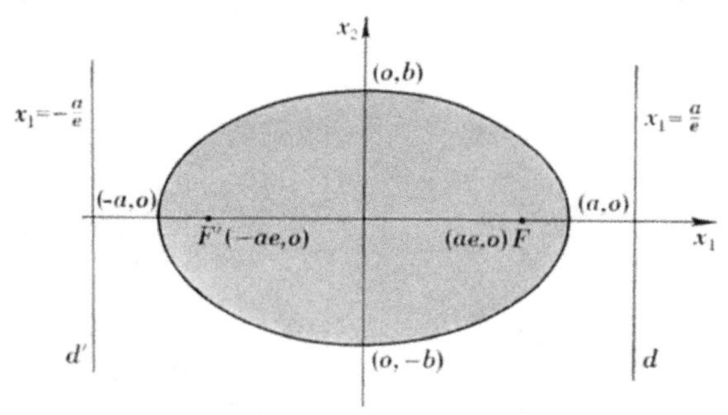

FIG. 6.2

In the case of the hyperbola there are two vertices $(\pm a,0)$ on the *transverse* axis, but the curve does not meet the *conjugate* axis $x_1 = 0$ in real points. The term "*principle axes*" is, however, still applicable.

If we think of an arbitrary point P on either an ellipse or an hyperbola and $\boldsymbol{\rho} = \overrightarrow{OP}$ as a radius vector, then the vertices may be defined in terms of the magnitude of $\boldsymbol{\rho}$. In the case of an ellipse, $\boldsymbol{\rho}$ takes its maximum value on the major axis and its minimum value on the minor axis, while in the case of an hyperbola $\boldsymbol{\rho}$ is a minimum on the transverse axis and has no maximum value.

There is one further conic called the *parabola*, whose equation in simplest form is

6.24
$$x_2^2 = 4px_1$$

The device we used before, of writing the equation in matrix form, is no longer applicable, but we shall see in a later chapter how our coordinate system can be modified to recover this convenience of expression.

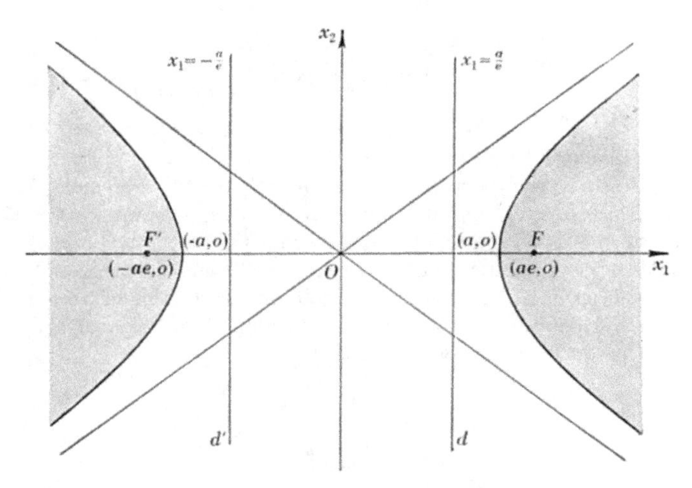

FIG. 6.3

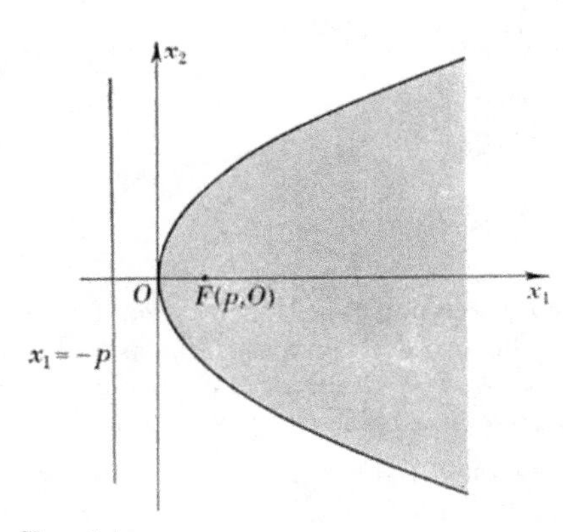

FIG. 6.4

Besides these analytical definitions, a conic can be defined synthetically. To this end we introduce the notion of the *eccentricity* e, where $e < 1$ for an ellipse, $e = 1$ for a parabola, and $e > 1$ for an hyperbola. The *foci* F, F' of an ellipse or an hyperbola have coordinates $(\pm ae,0)$ and these, along with corresponding *directrices* d, d', with equations $x_1 = \pm a/e$, are indicated in Figures 6.2 and 6.3. In the case of the parabola there is only one focus $(p,0)$ and one directrix $x_1 = -p$, as in Figure 6.4.

EXERCISES

1. Derive the equation 6.24 of the parabola as the locus of points P equidistant from the focus F and the directrix d.

2. Find the equation of the parabola with focus the origin and directrix the line $x_1 + x_2 = 1$.

3. Derive the equation 6.21 of the ellipse as the locus of points P such that PF is e times the perpendicular distance from P to the directrix d. Express b in terms of a and e. For what value of e is the locus a circle?

4. Show that the locus of Exercise 3 becomes the hyperbola 6.22 for $e > 1$.

5. Find the equation of an hyperbola with focus (1,1) and corresponding directrix $x_1 + x_2 = 1$ for which $e = 2$.

6. Write the equation of a conic through the five points (1,0), (0,1), (1,1), (−2,0), (0,−2) in determinantal form and expand to obtain the equation $x_1^2 - 2x_1x_2 + x_2^2 + x_1 + x_2 - 2 = 0$.

7. Prove that a conic is uniquely determined when *five* points on it are given.

6.3 QUADRICS AND THE LINES ON THEM

If we translate the origin to the point (a_1,a_2,a_3), the sphere 6.12 becomes

6.31 $$x_1^2 + x_2^2 + x_3^2 = r^2$$

The *normal form* of the equation of the *ellipsoid* (Figure 6.5) is

6.32 $$\frac{x_1^2}{a^2} + \frac{x_2^2}{b^2} + \frac{x_3^2}{c^2} = 1$$

while that of the *hyperboloid of one sheet* (Figure 6.6) is

6.33 $$\frac{x_1^2}{a^2} + \frac{x_2^2}{b^2} - \frac{x_3^2}{c^2} = 1$$

and that of the *hyperboloid of two sheets* (Figure 6.7) is

16.34 $$\frac{x_1^2}{a^2} + \frac{x_2^2}{b^2} - \frac{x_3^2}{c^2} = -$$

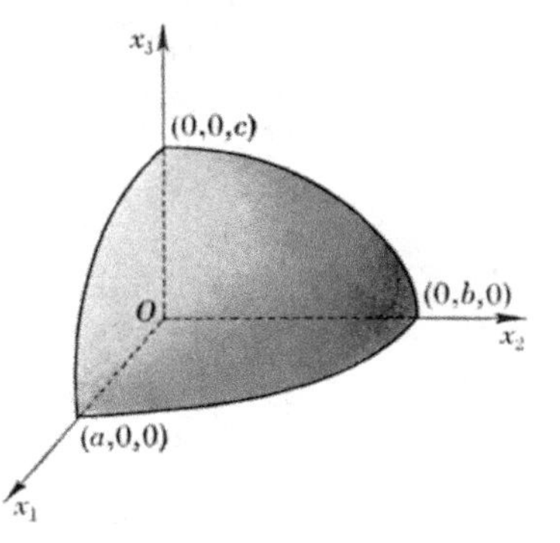

FIG. 6.5

Each of these *quadric surfaces* is symmetrical with respect to each coordinate plane, and so with respect to the origin which is here the *center* of the quadric.

The matrix forms of the equations analogous to 6.23 are:

6.321 $$\mathbf{X}^t \begin{pmatrix} \frac{1}{a^2} & 0 & 0 \\ 0 & \frac{1}{b^2} & 0 \\ 0 & 0 & \frac{1}{c^2} \end{pmatrix} \mathbf{X} = 1$$

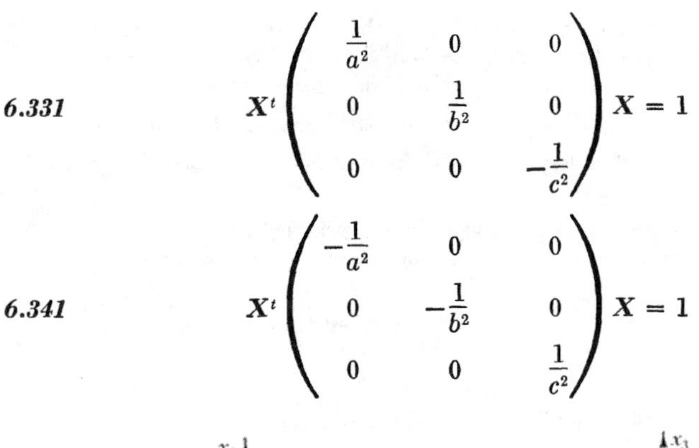

6.331 $$X^t \begin{pmatrix} \frac{1}{a^2} & 0 & 0 \\ 0 & \frac{1}{b^2} & 0 \\ 0 & 0 & -\frac{1}{c^2} \end{pmatrix} X = 1$$

6.341 $$X^t \begin{pmatrix} -\frac{1}{a^2} & 0 & 0 \\ 0 & -\frac{1}{b^2} & 0 \\ 0 & 0 & \frac{1}{c^2} \end{pmatrix} X = 1$$

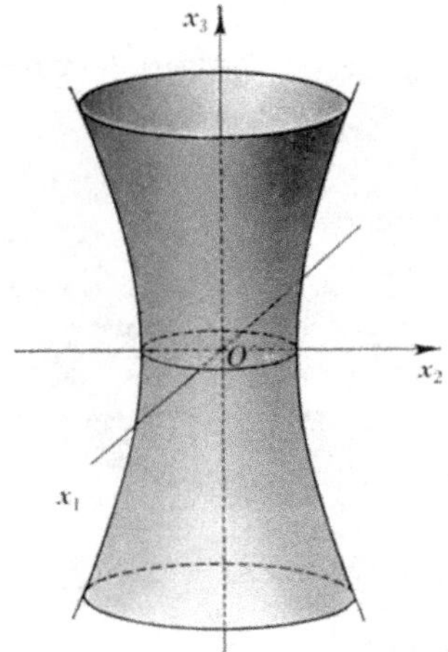

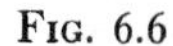
FIG. 6.6

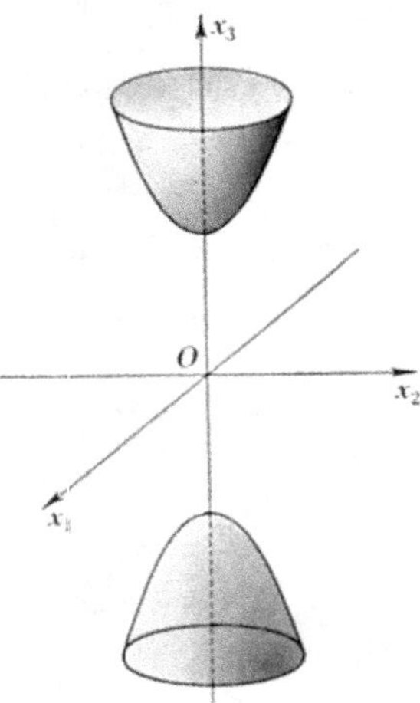

FIG. 6.7

By setting each coordinate equal to zero in turn, i.e., by taking the intersection of the surface with the corresponding coordinate plane, we may determine the shape of the surface. There are two noncentral quadrics: the *elliptic paraboloid* (Figure 6.8) with normal equation

6.35 $$\frac{x_1^2}{a^2} + \frac{x_2^2}{b^2} = x_3$$

and the *hyperbolic paraboloid* (Figure 6.9) with equation

6.36 $$\frac{x_1^2}{a^2} - \frac{x_2^2}{b^2} = x_3$$

Again we have no matrix forms of these equations until we suitably modify our coordinate system.

There is a remarkable property of the hyperboloid of one sheet 6.33

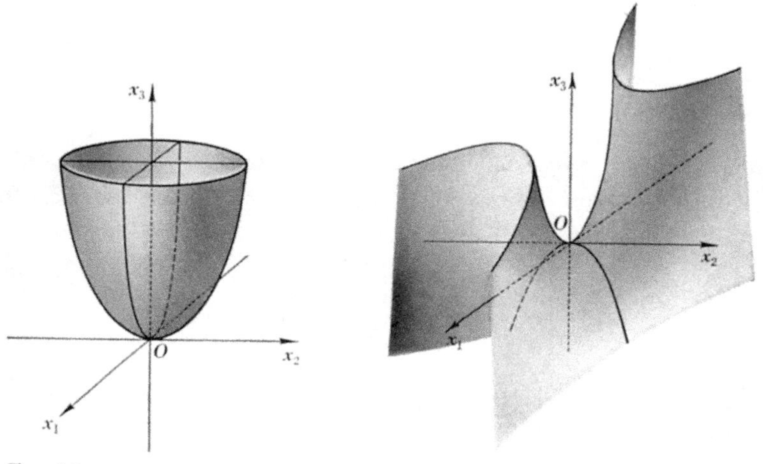

FIG. 6.8

FIG. 6.9

which is also shared by the hyperbolic paraboloid. Let us consider first the surface 6.33, writing the equation in the form

6.331
$$\frac{x_1^2}{a^2} - \frac{x_3^2}{c^2} = 1 - \frac{x_2^2}{b^2}$$

Factoring both sides we see that every point on the surface lies on a line

6.332
$$\frac{x_1}{a} + \frac{x_3}{c} = \lambda\left(1 + \frac{x_2}{b}\right), \qquad \frac{x_1}{a} - \frac{x_3}{c} = \frac{1}{\lambda}\left(1 - \frac{x_2}{b}\right)$$

and also on a line

6.333
$$\frac{x_1}{a} + \frac{x_3}{c} = \frac{1}{\mu}\left(1 - \frac{x_2}{b}\right), \qquad \frac{x_1}{a} - \frac{x_3}{c} = \mu\left(1 + \frac{x_2}{b}\right)$$

Conversely, every point on each of these lines lies in the surface, so that the surface is *ruled*. We call the lines 6.332 λ-*generators* and the lines 6.333 μ-*generators*, and prove two interesting theorems:

6.37 *Every λ-generator meets every μ-generator.*

Proof. If we write the determinant of the coefficients of x_1/a, x_2/b, x_3/c, 1 in the four equations 6.332 and 6.333, we must show that

$$\begin{vmatrix} 1 & -\lambda & 1 & \lambda \\ 1 & \dfrac{1}{\lambda} & -1 & \dfrac{1}{\lambda} \\ 1 & \dfrac{1}{\mu} & 1 & \dfrac{1}{\mu} \\ 1 & -\mu & -1 & \mu \end{vmatrix} = 0$$

and this can easily be verified.

6.38 *No two generators of the same family intersect.*

Proof. If we set $\lambda = \lambda_1, \lambda_2$ ($\lambda_1 \neq \lambda_2$) in 6.332, the result follows from the fact that

$$\begin{vmatrix} 1 & -\lambda_1 & 1 & \lambda_1 \\ 1 & \dfrac{1}{\lambda_1} & -1 & \dfrac{1}{\lambda_1} \\ 1 & -\lambda_2 & 1 & \lambda_2 \\ 1 & \dfrac{1}{\lambda_2} & -1 & \dfrac{1}{\lambda_2} \end{vmatrix} \neq 0$$

so that the corresponding equations are inconsistent.

These two properties hold also for lines of the two families

6.361 $$\frac{x_1}{a} + \frac{x_2}{b} = \lambda x_3, \qquad \frac{x_1}{a} - \frac{x_2}{b} = \frac{1}{\lambda}$$

6.362 $$\frac{x_1}{a} + \frac{x_2}{b} = \frac{1}{\mu}, \qquad \frac{x_1}{a} - \frac{x_2}{b} = \mu x_3$$

on the hyperbolic paraboloid 6.36, since the appropriate determinants can readily be constructed and evaluated.

If we take a fixed point P on, say, the hyperboloid of one sheet 6.33, then the two generators through P will define a plane which will be obtained by identifying the two equations

6.391 $$k_1\left\{\frac{x_1}{a} + \frac{x_3}{c} - \lambda\left(1 + \frac{x_2}{b}\right)\right\} + k_2\left\{\frac{x_1}{a} - \frac{x_3}{c} - \frac{1}{\lambda}\left(1 - \frac{x_2}{b}\right)\right\} = 0$$

6.392 $$l_1\left\{\frac{x_1}{a} - \frac{x_3}{c} - \mu\left(1 + \frac{x_2}{b}\right)\right\} + l_2\left\{\frac{x_1}{a} + \frac{x_3}{c} - \frac{1}{\mu}\left(1 - \frac{x_2}{b}\right)\right\} = 0$$

This identification yields $k_2 = \lambda = l_1$, $k_1 = \mu = l_2$, so that the equation of the plane in question is

6.393 $$\frac{x_1}{a}(\lambda + \mu) + \frac{x_2}{b}(1 - \lambda\mu) - \frac{x_3}{c}(\lambda - \mu) = 1 + \lambda\mu$$

where λ, μ are the parameters associated with the two generators of the surface through P. The plane so defined is the *tangent plane* at P, which can also be obtained by using the calculus, as we shall explain in the Appendix.

EXERCISES

1. Find the equations of the two families of generators of the hyperbolic paraboloid $x_1^2 - x_2^2 = x_3$, and in particular the equations of these generators through the point (2,1,3) on the surface.
2. Find the equation of the tangent plane to the surface at the point (2,1,3) in Exercise 1.
3. Find the equation of the tangent plane defined by a λ-generator and a μ-generator of the hyperbolic paraboloid 6.36, corresponding to the equation 6.393.
4. Solve the equations 6.332 and 6.333 to obtain the coordinates

$$\left(\frac{a(\lambda + \mu)}{1 + \lambda\mu}, \quad \frac{b(1 - \lambda\mu)}{1 + \lambda\mu}, \quad \frac{c(\lambda - \mu)}{1 + \lambda\mu}\right)$$

of the point of intersection of a λ-generator and a μ-generator of the hyperboloid 6.331. Verify that this point lies on the tangent plane 6.393. Explain why $1 + \lambda\mu \neq 0$.

6.4 CONES, CYLINDERS, AND SURFACES OF REVOLUTION

What locus is represented by the equation 6.11 in 3-space? Clearly x_3 is unrestricted, so that any point P whose first two coordinates satisfy 6.11 lies on the surface; thus any point on the line through P parallel to Ox_3 lies on the surface. This is the simplest example of a *right cylinder;* on such a surface there is one family of generators, each generator being orthogonal to the x_1x_2 plane. It is easy to see that any plane curve defines such a cylinder whose equation is that of the plane curve considered as a locus in 3-space.

A more interesting surface than the cylinder is the *cone.* For example, the locus represented by the quadratic equation

6.41 $$x_1^2 + x_2^2 = x_3^2$$

has the properties that (i) the origin lies on it; (ii) if (x_1, x_2, x_3) satisfies 6.41 then so also does $(\lambda x_1, \lambda x_2, \lambda x_3)$ for every λ; and also, every section of the surface by a plane $x_3 = k$ yields a circle. The surface represented by 6.41 is called a *right circular cone.* Indeed, the argument we have just given proves that

6.42 *Every homogeneous equation in x_1, x_2, x_3 represents a cone with vertex the origin.*

We can readily deduce other interesting properties of the right circular cone 6.41. If we take its intersection with the plane $x_1 = k$, we obtain

$$x_3^2 - x_2^2 = k^2$$

which is a *rectangular hyperbola* ($a = b = k$ in 6.22) with transverse axis parallel to Ox_3. If the secant plane intersects only that part of the cone above the x_1x_2 plane we have an *ellipse*, and if it is parallel to a generator, a *parabola*, as in Figure 6.10. It is these properties which led the Greeks to call these curves *conic sections* or *conics.*

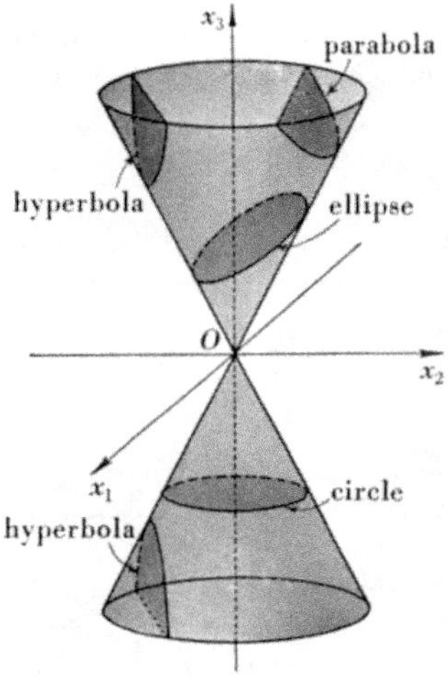

FIG. 6.10

A cylinder is a special case of a cone with vertex 'at infinity', and an arbitrary plane will intersect a circular cylinder in an ellipse, a circle, or a pair of parallel lines which may coincide. Of course the plane may not intersect the cylinder at all. In an exactly analogous manner a secant plane through the vertex O of the cone 6.41 may not intersect the cone in any other point, or if it does, the intersection will consist of two intersecting lines through O which again may coincide.

But we may also look at the right circular cylinder represented by the equation

6.43 $$x_1^2 + x_2^2 = r^2$$

and the right circular cone represented by the equation 6.41 as *surfaces of revolution.* In order to study such surfaces in general, consider the plane curve represented by the equation $f(x_2,x_3) = 0$ and imagine it rotated about the axis Ox_3 as in Figure 6.11. If we replace x_2 by $\sqrt{x_1^2 + x_2^2}$ in $f(x_2,x_3) = 0$ and rationalize, we have the desired equation.

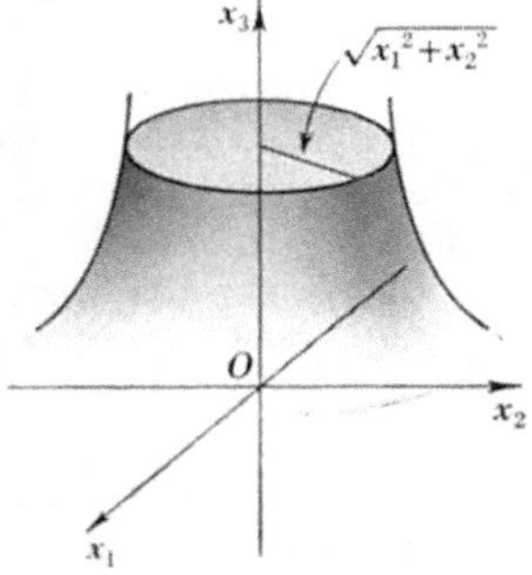

FIG. 6.11

Example. The line $x_2 = r$ when rotated yields the equation 6.43 and the line $x_2 = x_3$ yields the equation 6.41.

The representation of a quadric cone with vertex the origin by a matrix equation is interesting and instructive. The equation of the most general such cone could be written in the form

6.44 $$ax_1^2 + bx_2^2 + cx_3^2 + 2hx_1x_2 + 2gx_1x_3 + 2fx_2x_3 = 0$$

which we can rewrite

6.45 $$\mathbf{X}^t \begin{pmatrix} a & h & g \\ h & b & f \\ g & f & c \end{pmatrix} \mathbf{X} = 0$$

The form of the matrix in 6.45 is significant since, denoting it by M, we have

6.46 $$M^t = M$$

and such a matrix is said to be *symmetric* (cf. Exercises 3–5 of Section 3.2). Conversely, any symmetric matrix yields a cone with vertex the origin, unless it degenerates (cf. Exercise 5 of Section 6.5).

EXERCISES

1. $$\frac{x_1^2}{2} + \frac{x_2}{3} = 1, \qquad \frac{x_1}{2} - \frac{x_2^2}{3} = 1$$
represent two conics in the plane $x_3 = 0$. Find the equations of the surfaces of revolution obtained by rotating each conic about each of its principle axes. Roughly sketch the four surfaces so obtained.

2. Find the equations of the cones formed by rotating the line $x_3 = 0$, $x_2 = 2x_1$ about (a) the x_1 axis, (b) the x_2 axis.

3. Which of the different types of quadric surface can be realized as surfaces of revolution? Make rough sketches to illustrate your answer.

4. Derive the equation of the *torus* (anchor ring) generated by rotating the circle
$$x_1^2 + x_2^2 + 2ax_1 + b^2 = 0 = x_3 \qquad (a^2 > b^2)$$
about the x_2 axis.

5. Discuss the intersection of the plane $x_3 = k$ with the torus in Exercise 3, for all values of k.

6.5 PAIRS OF LINES AND PLANES

We may well ask when a quadratic equation in two variables,

6.51 $$ax_1^2 + bx_2^2 + 2hx_1x_2 + 2gx_1 + 2fx_2 + c = 0$$

factorizes, and so represents a pair of lines. One may approach the problem directly and identify the coefficients of the equation 6.51 with those of the equation

$$(lx_1 + mx_2 + n)(l'x_1 + m'x_2 + n') = 0 \tag{6.52}$$

to obtain

$$ll' = a, \quad mm' = b, \quad nn' = c$$

$$lm' + l'm = 2h, \quad ln' + l'n = 2g, \quad mn' + m'n = 2f$$

Substituting,

$$\begin{aligned} 8fgh &= 2ll'mm'nn' + ll'(m'^2n^2 + m^2n'^2) \\ &\quad + mm'(n'^2l^2 + n^2l'^2) + nn'(l'^2m^2 + l^2m'^2) \\ &= 2abc + a(4f^2 - 2bc) + b(4g^2 - 2ca) + c(4h^2 - 2ab) \end{aligned}$$

Collecting terms and dividing out the factor 4 we obtain as the required condition:

$$\begin{vmatrix} a & h & g \\ h & b & f \\ g & f & c \end{vmatrix} = 0 \tag{6.53}$$

A more significant approach to the condition 6.53 uses the calculus, as will be explained in the Appendix.

The curve represented by the general equation 6.51 is met by an arbitrary line in two points which may coincide, or in no points at all. This follows immediately by eliminating one of the variables in 6.52 by substituting from the general linear equation

$$u_1x_1 + u_2x_2 + u_0 = 0 \tag{6.54}$$

and considering the possible roots of the resulting quadratic equation. Alternatively, we may suppose the line 6.54 to be defined parametrically by equations

$$x_1 = y_1 + l_1t$$

$$x_2 = y_2 + l_2t$$

as in Section 1.2. Substituting in 6.51 we have a quadratic equation in t and the same argument applies.

It would be possible to study the effect of a linear transformation on the equation 6.51 and show that by suitably choosing the constants we could bring it into one of the normal forms 6.21, 6.22, 6.24 if it did not factorize into a product of linear factors. A similar procedure is applicable to the general quadratic equation in x_1, x_2, x_3 and we conclude that every quadric is reducible to one of the normal forms 6.32–6.36 or is recognizable immediately as a cone, a cylinder, or a pair of planes. The method is the same in each case but we shall postpone its consideration to the final chapter of this book.

Just as in the case of a cone, *an arbitrary plane π meets any quadric Q in a conic C or a pair of lines.* This follows from the fact that two planes define a line and by elimination we conclude that any line in π meets Q, and so C, in two points which may coincide or in no points at all. The locus C is quadratic and so must be a conic or a pair of lines.

EXERCISES

1. Obtain the condition 6.53 by solving 6.51 as a quadratic in x_1, and insisting that the discriminant be a perfect square.
2. For what values of k does the quadratic equation

$$2x_1^2 - x_2^2 - x_1x_2 + (2 - k)x_1 + (1 + k)x_2 = k$$

represent a pair of lines? Find the lines.

3. Test the quadratic equation

$$x_1^2 + x_1x_2 - 2x_2^2 + 3x_2 = 1$$

to see if it factorizes, and if so, obtain the factors. Plot the locus on a sheet of graph paper.

4. Write each of the equations

$$x_1^2 - x_1x_2 + x_2^2 - x_3^2 = 0$$

$$2x_1^2 - x_2^2 - x_3^2 - x_1x_2 + x_1x_3 + 2x_2x_3 = 0$$

$$(x_1 - x_2 + 2x_3)^2 = 0$$

in matrix form 6.45.

5. What is the significance of the condition 6.53 for the equation of a cone 6.45? Test your answer on the equations of Exercise 4, and describe the loci.
6. Prove that the equation

$$\mathbf{X}'\begin{pmatrix} a & h & g \\ h & b & f \\ g & f & c \end{pmatrix}\mathbf{X} = 1$$

represents a central quadric.

6.6 A QUADRIC TO CONTAIN THREE SKEW LINES

In Section 6.3 we saw that a hyperboloid of one sheet or a hyperbolic paraboloid has on it two families of generators with the properties 6.37

and 6.38. One is tempted to ask the question: Given three skew lines in space is there a quadric containing them?

To be specific, let us take the three lines l_1, l_2, l_3 to be three non-intersecting edges of a rectangular parallelepiped with center the origin and equations as indicated in Figure 6.12 below. The most general quadric surface to contain l_1 and l_2 would have the form

6.61 $$\alpha(x_1 - a_1)(x_1 + a_1) + \beta(x_1 - a_1)(x_3 - a_3) + \gamma(x_1 + a_1)(x_2 + a_2) + \delta(x_2 + a_2)(x_3 - a_3) = 0$$

and if this is to be satisfied by *every* point of l_3 then we must have

$$\begin{aligned} 0 &= \alpha(x_1 - a_1)(x_1 + a_1) + \beta(x_1 - a_1)(-2a_3) \\ &\quad + \gamma(x_1 + a_1)(2a_2) + \delta(2a_2)(-2a_3) \\ &= \alpha x_1^2 + 2(-a_3\beta + a_2\gamma)x_1 + (-a_1^2\alpha + 2a_1a_3\beta + 2a_1a_2\gamma - 4a_2a_3\delta) \end{aligned}$$

so that

$$\alpha = 0, \quad a_3\beta = a_2\gamma, \quad a_1\gamma = a_3\delta$$

from which we conclude that

6.62 $$\alpha = 0, \quad \frac{\beta}{a_2} = \frac{\gamma}{a_3} = \frac{\delta}{a_1}$$

and 6.61 takes the simple form

6.63 $$a_1x_2x_3 + a_2x_1x_3 + a_3x_1x_2 + a_1a_2a_3 = 0$$

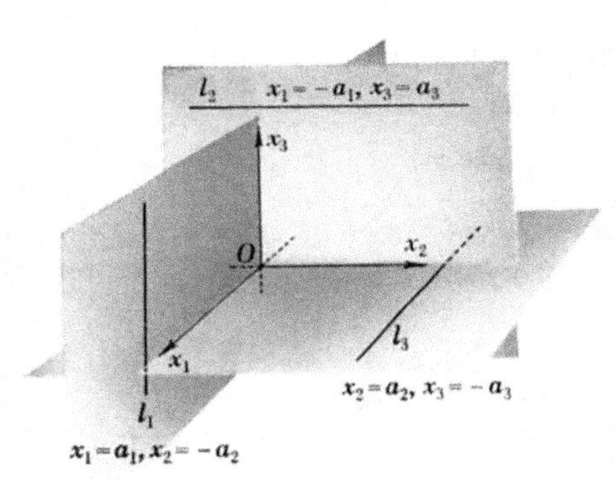

FIG. 6.12

While the choice of the lines l_1, l_2, l_3 would seem specialized, no metrical property is involved and we have only chosen them symmetrically with regard to the basis vectors and this can always be arranged. That the ruled surface in question is uniquely defined is important and since it is centrally symmetrical it must be a hyperboloid of one sheet. If the three lines are parallel to the same plane, we no longer have such central symmetry and the surface is an hyperbolic paraboloid.

Inserting the values of α, β, γ, δ from 6.62, the equation 6.61 takes the form

6.64 $$a_1(x_2 + a_2)(x_3 - a_3) + a_2(x_1 - a_1)(x_3 - a_3) + a_3(x_1 + a_1)(x_2 + a_2) = 0$$

which factors into

$$(x_2 + a_2) = \lambda(x_1 - a_1)$$

6.65

$$a_1(x_3 - a_3) + a_3(x_1 + a_1) = -\frac{a_2}{\lambda}(x_3 - a_3)$$

and similarly into

$$(x_2 + a_2) = \mu(x_3 - a_3)$$

6.66

$$a_1(x_3 - a_3) + a_3(x_1 + a_1) = -\frac{a_2}{\mu}(x_1 - a_1)$$

if we set $\mu = 0, \infty, -a_2/a_3$, we obtain the lines l_1, l_2, l_3 respectively. On the other hand, we may argue as in Section 6.3, that every line 6.65 meets every line 6.66 since the determinant of the coefficients vanishes identically in λ and μ.

We may look at the two systems of generators on a hyperboloid as yielding a parametric representation of the surface in the sense that each point P is the intersection of a unique λ-generator and a unique μ-generator (cf. Exercise 4 of Section 6.3). Alternatively, we may use trigonometric functions. Just as the ellipse 6.21,

$$\frac{x_1^2}{a^2} + \frac{x_2^2}{b^2} = 1$$

is represented parametrically in the form $x_1 = a \cos \theta$, $x_2 = b \sin \theta$, so the ellipsoid 6.32,

$$\frac{x_1^2}{a^2} + \frac{x_2^2}{b^2} + \frac{x_3^2}{c^2} = 1$$

is represented parametrically in the form

6.67 $$x_1 = a \cos \theta \sin \varphi, \quad x_2 = b \sin \theta \sin \varphi, \quad x_3 = c \cos \varphi$$

Corresponding representations can be constructed for the other quadrics.

EXERCISES

1. Construct a parametrization for the hyperboloid 6.33 corresponding to 6.67 for the ellipsoid.

2. Verify that the hyperbolic paraboloid 6.36 can be represented parametrically by the equations

$$x_1 = ae^{\varphi} \cosh \theta, \quad x_2 = be^{\varphi} \sinh \theta, \quad x_3 = e^{2\varphi}$$

and that $\theta + \varphi$ is constant for points of a given generator of one system and $\theta - \varphi$ is constant for a given generator of the other system.

3. The equation of an ellipsoid 6.32 may be written

$$\frac{x_1^2 + x_2^2 + x_3^2}{b^2} - 1 + x_1^2\left(\frac{1}{a^2} - \frac{1}{b^2}\right) + x_3^2\left(\frac{1}{c^2} - \frac{1}{b^2}\right) = 0$$

If $a > b > c$, show that

$$x_1^2\left(\frac{1}{a^2} - \frac{1}{b^2}\right) + x_3^2\left(\frac{1}{c^2} - \frac{1}{b^2}\right) = 0$$

represents two planes which intersect the ellipsoid in circles of radius b.

4. Apply the method of Exercise 3 to find the circular sections of the hyperboloid.

5. If we denote the matrix in 6.45 by M, we may write the equation of Exercise 6 of Section 6.5 in the form

$$\mathbf{X}'M\mathbf{X} - \lambda\mathbf{X}'\mathbf{X} + \lambda\left(\mathbf{X}'\mathbf{X} - \frac{1}{\lambda}\right) = 0$$

For what values of λ does the equation

$$\mathbf{X}'M\mathbf{X} - \lambda\mathbf{X}'\mathbf{X} = 0$$

represent a pair of planes? Discuss the intersections of the planes with the quadric.

6.7 THE INTERSECTION OF TWO QUADRICS

Finally, we consider the nature of the intersection of two quadric surfaces Q_1, Q_2. The simplest procedure is to consider the intersection of each surface with a given plane π, so that we have two conics C_1, C_2 in π. If C_1 and C_2 intersect, their common points belong to both Q_1 and Q_2. Since C_1 and C_2 intersect in at most 4 points, we say that the curve of intersection of Q_1 and Q_2 in space is of *order* 4. In particular, this *twisted* quartic curve may degenerate into a common generator and a twisted cubic curve. We investigate this case briefly.

Before doing so, however, we remark that a *curve C of the second order in space must be a conic*, i.e., it must lie in a plane; since, if not, a plane containing two points of C could be chosen so as to meet it in a third point and C would have order greater than 2.

It is sufficient to consider the simplest case, e.g., the intersection of the cone

6.71 $$x_2^2 = x_1x_3$$

with the hyperbolic paraboloid

6.72 $$x_1x_2 = x_3$$

The generators of the cone 6.71 are given by the equations

6.73 $$x_2 = \lambda x_1, \qquad x_2 = \frac{1}{\lambda}x_3$$

and those of the hyperbolic paraboloid 6.72 by the equations

6.74
$$x_1 = \lambda, \quad x_2 = \frac{1}{\lambda} x_3; \qquad x_2 = \mu, \quad x_1 = \frac{1}{\mu} x_3$$

and it is easy to see that the two surfaces have in common the x_1 axis, i.e., the line

6.75
$$x_2 = 0 = x_3$$

The residual intersection is the twisted cubic curve given parametrically by the equations

6.76
$$x_1 = \theta, \quad x_2 = \theta^2, \quad x_3 = \theta^3$$

EXERCISE

Make a sketch of the two surfaces 6.71 and 6.72 and indicate as best you can the nature of the intersection.

7 HOMOGENEOUS COORDINATES AND PROJECTIVE GEOMETRY

7.1 EUCLIDEAN GEOMETRY

So far, we have accepted the two basic features of Euclidean geometry without question: namely, *parallelism* and the *Pythagorean theorem.* These are not unrelated, for a fundamental property of two parallel lines is that they are *equidistant,* and the notion of distance is defined by means of the Pythagorean theorem. Even if we agreed not to use the Pythagorean theorem, however, we could still speak of parallel lines using Euclid's definition that there is a unique line through a given point P, coplanar with a given line l and not meeting it.

But we may look at the matter differently. Consider a Euclidean plane π and a point O not in π. Every point P in π determines a unique line OP and every line l in π a unique plane Ol. Two lines l, l' in π which intersect in P determine two planes Ol, Ol' which intersect in OP; while if l and l' are parallel in π then the planes Ol, Ol' still intersect in a unique line OP_∞ parallel to π. All such lines OP_∞ lie in the plane π_∞ through O, parallel to π.

Let us set out this correspondence between the points and lines of π and the lines and planes of the *bundle* with vertex O, in the following fashion:

Point P of $\pi \leftrightarrow$ Line OP

Line l of $\pi \leftrightarrow$ Plane Ol

Two points P, Q of π determine a line PQ of π.	$\leftrightarrow$	Two lines OP, OQ determine a plane OPQ.
Two lines l, l' of π either intersect or are parallel.	$\leftrightarrow$	Two planes Ol, Ol' always intersect.

If we associate a line OP through O with a point P in π when OP intersects π in P, why should we not similarly associate a point P_∞ *at infinity* in π with the line OP_∞? Indeed this association is well defined, since any plane Ol' will contain OP_∞ if and only if l' is parallel to l. Moreover, all lines OP_∞ lie in the plane π_∞ parallel to π, so that it is natural to speak of P_∞ as lying in the line l_∞ "at ∞" in π.

In the following section we shall see how this can all be done analytically.

7.2 HOMOGENEOUS COORDINATES

Let us reconsider the problem of finding the common point of two coplanar lines. Assuming now that these lines are parallel, we may take their equations in the form

7.211
$$\begin{aligned} a_{11}x_1 + a_{12}x_2 &= a_{10} \\ a_{11}x_1 + a_{12}x_2 &= a_{20} \end{aligned}$$

Previously we said that such equations are inconsistent for $a_{10} \neq a_{20}$ and have no solution. If, however, we introduce a new "variable of homogeneity" x_0 and write

7.212
$$\begin{aligned} a_{11}x_1 + a_{12}x_2 &= a_{10}x_0 \\ a_{11}x_1 + a_{12}x_2 &= a_{20}x_0 \end{aligned}$$

then the two equations do have a solution $x_1 = ka_{12}$, $x_2 = -ka_{11}$, $x_0 = 0$ for all values of k. If we call x_1, x_2, x_0 the *homogeneous coordinates* of the common point, the two approaches can be reconciled by writing the familiar nonhomogeneous coordinates, which we temporarily denote $\bar{x}_1$, $\bar{x}_2$, $\bar{x}_3$, . . . , in the form

$$7.22 \qquad \bar{x}_1 = \frac{x_1}{x_0}, \quad \bar{x}_2 = \frac{x_2}{x_0}, \quad \bar{x}_3 = \frac{x_3}{x_0}, \quad \ldots$$

Notice that the homogeneous coordinates $x_1, x_2, x_3, \ldots x_0$ are determined up to a constant factor $k \neq 0$ only. For any finite point we may set $x_0 = 1$ so that $\bar{x}_i = x_i$ $(i = 1,2,3, \ldots)$ but for points "*at infinity*," $x_0 = 0$, which is the equation of the *space at infinity*. The homogeneous coordinates of the origin can be taken to be $(0,0,0, \ldots, 1)$ while those of the point *at infinity* on Ox_1 can be taken to be $(1,0,0, \ldots, 0)$, those of the point at infinity on Ox_2 can be taken to be $(0,1,0, \ldots, 0)$, and so on. There is no ambiguity here since division by zero is not allowed, and multiplication by any $k \neq 0$ does not change the point so represented.

Just as the equation of a line in two dimensions can be taken to be homogeneous as in 7.212, so the equation of a plane in three dimensions can be taken to be homogeneous:

$$7.23 \qquad a_1x_1 + a_2x_2 + a_3x_3 = a_0x_0$$

We may consider the arbitrary constant factor k, by which the equation may be multiplied without changing its geometrical significance, to be precisely that k for which the point

$$(x_1,x_2,x_3,x_0) = (kx_1,kx_2,kx_3,kx_0)$$

Thus we have made the designation of a *point* in space conform to the same algebraic convention as holds for the *equation* of any locus.

If we write the general equation of a conic in the form

$$7.241 \qquad a\bar{x}_1^2 + b\bar{x}_2^2 + 2h\bar{x}_1\bar{x}_2 + 2g\bar{x}_1 + 2f\bar{x}_2 + c = 0$$

it assumes the homogeneous form

$$7.242 \qquad ax_1^2 + bx_2^2 + 2hx_1x_2 + 2gx_1x_0 + 2fx_2x_0 + cx_0^2 = 0$$

when we replace nonhomogeneous by homogeneous coordinates. Similarly, any algebraic equation* can be made homogeneous and, indeed, equations *which are nonhomogeneous in* x_1, x_2, x_0 *have no geometrical significance.*

On introduction of the coordinate of homogeneity x_0, the set of numbers (x_1,x_2,x_0) can no longer be considered as a vector in the plane. Nevertheless, there is the considerable advantage that the quadratic equation 7.242 can now be written

$$7.243 \qquad (x_1, x_2, x_0)\begin{pmatrix} a & h & g \\ h & b & f \\ g & f & c \end{pmatrix}\begin{pmatrix} x_1 \\ x_2 \\ x_0 \end{pmatrix} = 0$$

* We assume that we have a *rational integral* algebraic function, i.e., a *polynomial*, set equal to zero. Such a polynomial is said to be *homogeneous of weight* w if, when every x_i is replaced by x_it, exactly t^w is a factor of every term.

yielding a significant generalization of the equations 6.23. The similarity of 7.243 and 6.45 is suggestive. Indeed, we obtain the nonhomogeneous equation of a conic by setting $x_3 = 1$ in 6.45, i.e., by taking the intersection of the cone by this plane. In so doing we have established exactly the correspondence envisaged in Section 7.1. The vector $\mathbf{X}(x_1,x_2,x_3)$ in 3-space defines a unique point (x_1,x_2,x_0) in the plane $x_3 = x_0 = 1$, and (kx_1,kx_2,kx_3) defines the same point for every $k \neq 0$.

In order to generalize these ideas further, it is convenient to write the equation 7.243 in the form

7.25
$$(x_1,\ x_2,\ x_0)\begin{pmatrix} a_{11} & a_{12} & a_{10} \\ a_{21} & a_{22} & a_{20} \\ a_{01} & a_{02} & a_{00} \end{pmatrix}\begin{pmatrix} x_1 \\ x_2 \\ x_0 \end{pmatrix} = 0$$

where $a_{ij} = a_{ji}$. In this notation the general homogeneous quadratic equation in x_1, x_2, x_3, x_0 becomes

7.26
$$(x_1,\ x_2,\ x_3,\ x_0)\begin{pmatrix} a_{11} & a_{12} & a_{13} & a_{10} \\ a_{21} & a_{22} & a_{23} & a_{20} \\ a_{31} & a_{32} & a_{33} & a_{30} \\ a_{01} & a_{02} & a_{03} & a_{00} \end{pmatrix}\begin{pmatrix} x_1 \\ x_2 \\ x_3 \\ x_0 \end{pmatrix} = 0$$

where again $a_{ij} = a_{ji}$. Such an equation contains as a special case each of 6.321 through 6.341, and 7.26 represents a general quadric surface. We shall see how to reduce 7.25 and 7.26 to normal form in the last chapter of this book.

EXERCISES

1. Write the equations

$$x_1 - 2x_2 = 3$$
$$2x_1 - 4x_2 = 2$$

in homogeneous form and solve.

2. What is the nonhomogeneous form of the equation of the curve

$$x_0x_1^2 - x_2(x_2^2 - x_0^2) = 0?$$

Find the homogeneous coordinates of its intersections with the axes and locate the points in question. Roughly sketch the curve.

3. Write the equation

$$x_1^2 - x_2^2 + x_1x_2 - 2x_1 + 3x_2 = 1$$

in the form 7.25 with *integral* coefficients.

4. Write the equation

$$x_1x_2 - x_1x_3 + x_2x_3 = 1$$

in the form 7.26 with integral coefficients. What difference would it have made if the right side had been 0 instead of 1?

7.3 AXIOMS OF PROJECTIVE GEOMETRY

By introducing homogeneous coordinates we have been able to extend our analytical machinery so as to take account of the behavior of geometric loci *at infinity*. In a very practical sense we have *adjoined* such points, and the line or plane (space) *at infinity* in which they lie, to the ordinary Euclidean plane or space with which we are familiar. We have, in fact, made it possible to say that *any* two coplanar lines have a point in common (finite or infinite) and *any* two planes in space have a line in common (finite or infinite) and *every* line meets *every* plane in a point (finite or infinite). If we do not involve the notion of *distance* or *length*, i.e., if our space has no *metric* imposed upon it (e.g., by a Pythagorean theorem), we have what is called *affine* geometry. If we do not distinguish between finite and infinite elements, we have *projective* geometry.

In order to clarify these ideas, let us approach the situation from the opposite point of view and give a system of *incidence axioms* which will define a projective space, say of three dimensions. To this end we take a *point* to be undefined and a *line* to be an undefined class of at least two points. It is important to be somewhat vague here so that our system of axioms may be capable of different interpretations. In this way we can include many apparently diverse systems which are subject to the same *relations*, when *point* and *line* are interpreted differently. Concerning these undefined elements we make the following assumptions:

7.31 There are at least two distinct points.

7.32 Two distinct points A, B determine one and only one line AB (or BA) through both A and B.

It is not difficult to prove that if C and D are points on AB, then A and B are points on CD. Moreover, two distinct lines cannot have more than one common point.

7.33 If A, B are distinct points, then there is at least one point C distinct from A, B on the line AB.

7.34 If A, B are distinct points, then there is at least one point C not on the line AB.

7.35 If A, B, C are Three noncollinear points and D is a point on BC distinct from B and C and E is a point on CA distinct from C and A, then there is a point F on AB such that D, E, F are collinear.

This axiom 7.35, first stated by Pasch in 1880, makes it possible to define a *plane ABC* and to prove that *any two coplanar lines have a point in common.* In order to have a space of at least three dimensions, we assume that:

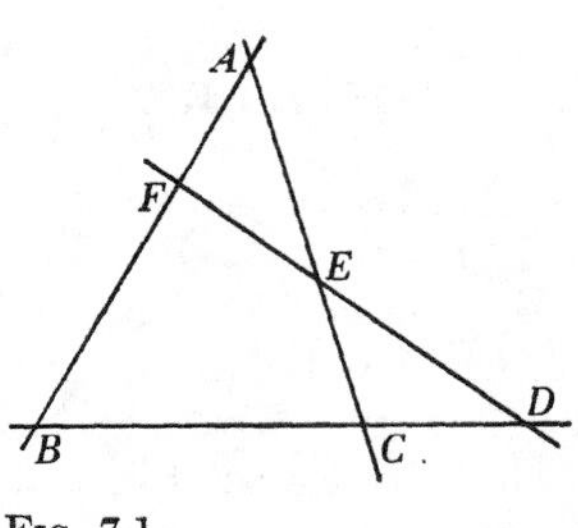

FIG. 7.1

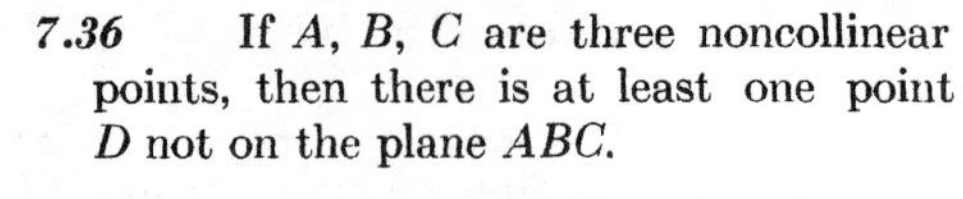

7.36 If A, B, C are three noncollinear points, then there is at least one point D not on the plane ABC.

To exclude the possibility that it have more than three dimensions, we assume finally that

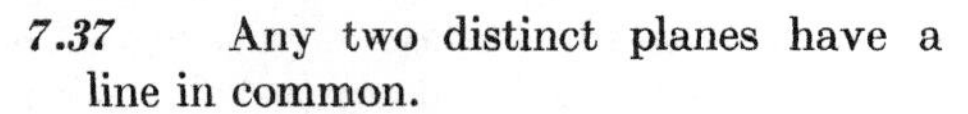

7.37 Any two distinct planes have a line in common.

These seven axioms describe how points, lines, and planes are *related* to each other in projective geometry. There must be at least three points on a line; how many more is not specified. Already we have an example of the utility of leaving "point" undefined. In fact, the interpretations of "point" as a "line of the bundle" and "line" as a "plane of the bundle" in Section 7.1 satisfy all our assumptions, and we conclude that the geometry of the bundle is in one-to-one correspondence with the geometry of the projection plane.

7.4 THEOREMS OF DESARGUES AND PAPPUS

Many beautiful theorems can be proved on the basis of these assumptions, but we prove only one due to the French geometer Desargues (1593–1662).

7.41 Theorem of Desargues *If two triangles ABC, $A'B'C'$ are situated in the same or in different planes and are such that BC, $B'C'$ meet in L, CA, $C'A'$ meet in M, and AB, $A'B'$ meet in N, where L, M, N are collinear, then AA', BB', CC' are concurrent, and conversely.*

Proof. (i) If we assume that the two triangles are in different planes π, π', then the three points L, M, N must lie on the line l common to π and π'. Since A, A', B, B' are coplanar, as also are B, B', C, C' and C, C', A, A', we know that these three planes must have a point O in common so that AA', BB', CC' all pass through O. The converse theorem follows by reversing the argument.

(ii) If the two triangles ABC, $A'B'C'$ lie in the same plane π, we may choose a plane π_1 through l distinct from π and a point P not in π or π_1. Projecting $A'B'C'$ from P into a triangle $A_1B_1C_1$ in π_1, we know the theorem is true for ABC and $A_1B_1C_1$, so that there exists a point O_1 in which AA_1, BB_1, CC_1 all meet. Projecting back again onto π from P, the point O_1 projects into a point O in which AA', BB', CC' all meet, as required. The converse follows similarly.

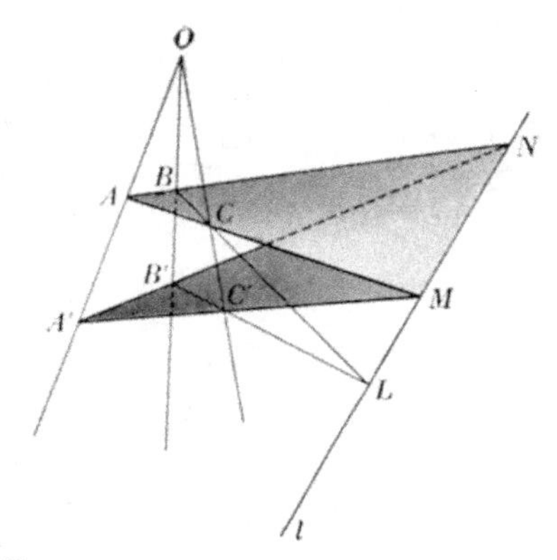

Fig. 7.2

The remarkable fact about Desargues' theorem is that it cannot be proved for coplanar triangles unless the plane containing them is embedded in a three-dimensional space. Examples of *non-Desarguesian* planes can be constructed which satisfy axioms 7.31–7.35 without Desargues' theorem being valid (cf. the exercise of Section 7.6). There is thus a notable difference between spaces of two and three or more dimensions which is far from being fully understood.

While the theorem of Pappus (3rd Century A.D.) is expressible in terms of incidence relations, it cannot be proved without further assumptions:

7.42 *Theorem of Pappus* *If A, B, C are any three distinct points on a line l and A', B', C' any three points on a line l' intersecting l, then the three points $(BC', B'C)$, $(CA', C'A)$, $(AB', A'B)$ are collinear.*

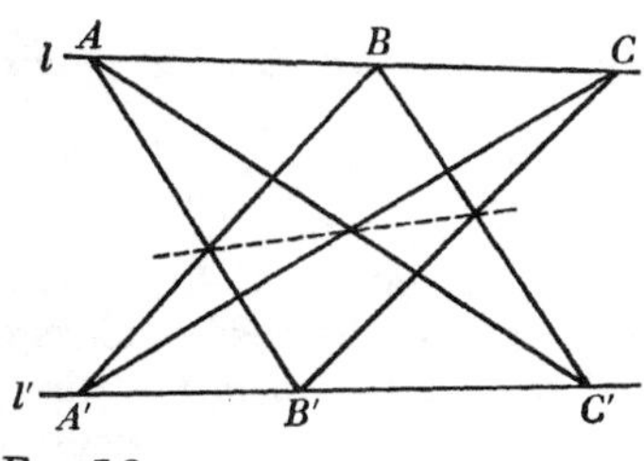

Fig. 7.3

Only relatively recently has the full significance of these theorems of Desargues and Pappus been brought to light in Hilbert's *Foundation of Geometry* (Chicago: Open Court, 1938).

EXERCISES

1. Formulate the theorem of Desargues in the bundle and verify that the proof remains valid.
2. Formulate the theorem of Pappus in the bundle.

7.5 AFFINE AND EUCLIDEAN GEOMETRY

Having defined a projective space, in particular a projective plane π, we can specialize an arbitrary line l_∞ in π which we may designate as the "*line at* ∞." Any two lines l, l' which intersect on l_∞ will be called *parallel*. If l meets l_∞ in P_∞, then through any point P not on l there passes one and only one line PP_∞ parallel to l, and this is Euclid's axiom of parallelism. Such a definition of parallelism involves no *metric*, i.e., no measure of length or of angle, and yields *affine* geometry.

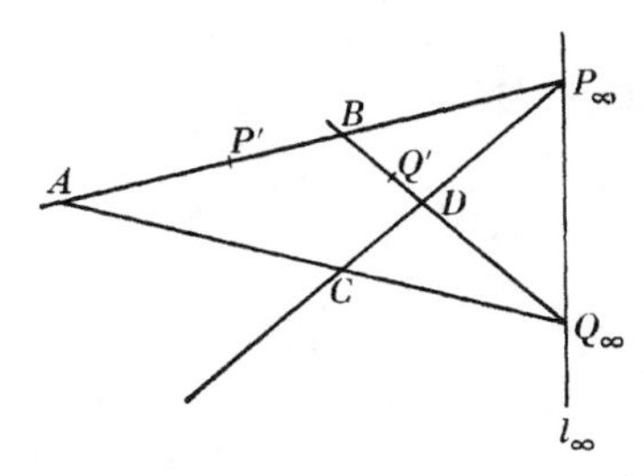

FIG. 7.4

In order to introduce a measure of length it is natural to proceed in two stages: (i) to define equality of segments under *parallel translation* and then (ii) under *rotation*.

(i) Having defined parallelism in affine geometry, we have a parallelogram $ABCD$ if $AB \| CD$ and $AC \| BD$, as in Figure 7.4. The statement that $AB = CD$ and $AC = BD$ is natural, and leads to all the familiar consequences.

(ii) In order to define equality of segments under *rotation*, we must distinguish a *circle* from the general conic 7.242. The most natural way of doing this is to observe that every circle

7.51 $$x_1^2 + x_2^2 + 2gx_1x_0 + 2fx_2x_0 + cx_0^2 = 0$$

intersects the line at infinity $x_0 = 0$ in the two so-called "*circular points at infinity*" given by the equations

7.52 $$x_1^2 + x_2^2 = 0 = x_0$$

These points have the conjugate complex coordinates $(1, \pm i, 0)$. Conversely, any conic through these two *circular points* is by definition a circle. By choosing two such points on l_∞ and designating them as "circular points," it is possible to introduce a measure of length and eventually the full Euclidean metric.

Though we cannot go into details here, this building up of Euclidean geometry from projective and affine geometry leads to a clearer understanding of the ideas involved. In the following chapter, we shall study geometry on the surface of a sphere. Besides being our "homeland" this provides the simplest example available of a *non-Euclidean* metric.

EXERCISES

1. Set up equations corresponding to 6.15 to represent the parallel translation of the origin to the point $P(p_1,p_2)$. What would these become in homogeneous coordinates?
2. Prove that the equation 7.243 of any conic passing through the circular points $(1, \pm i, 0)$ must reduce to the equation 7.51 of a circle.
3. What is the effect of the parallel translation of Exercise 1 on the circular points?
4. Write the general linear transformation in the plane in homogeneous coordinates and consider its effect on the line at infinity. What is the condition that it leave l_∞ (i) invariant, (ii) pointwise invariant? In which of these categories would you place (a) a parallel translation, (b) a rotation about the origin?
5. Write the general linear transformations of Exercise 4 in nonhomogeneous form. What would be the corresponding transformations in space?

7.6 DESARGUES' THEOREM IN THE EUCLIDEAN PLANE

In order to clarify these ideas still further, let us consider the theorem of Desargues from the point of view of Euclidean geometry. In Figure 7.2 of Section 7.4 we could assume that all the intersections are "finite" and in this case the proof given is applicable in the Euclidean case. Let us consider in particular the two cases illustrated below in which one (or two and so each) side of the triangle $A'B'C'$ is parallel to the corresponding side of the triangle ABC. Both triangles are here assumed to lie in the same plane.

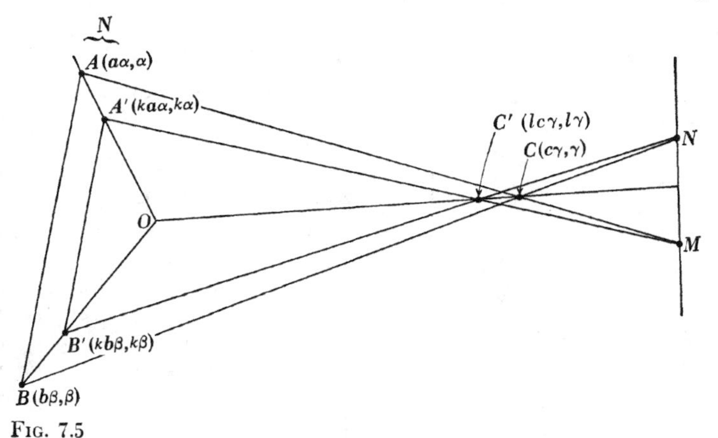

FIG. 7.5

Case (i). We take the center of perspective O as origin and the coordinates and equations as indicated in Figure 7.5. The parallelism of AB and $A'B'$ is expressed by introducing the constant $k \neq 0$ in the manner indicated; we assume $l \neq k$. Writing the equations of AC and $A'C'$ in parametric form, we have

7.61 $$AC : \begin{cases} x_1 = a\alpha + (a\alpha - c\gamma)s \\ x_2 = \alpha + (\alpha - \gamma)s \end{cases}$$

and

7.62 $$A'C' : \begin{cases} x_1 = ka\alpha + (ka\alpha - lc\gamma)t \\ x_2 = k\alpha + (k\alpha - l\gamma)t \end{cases}$$

and if these are identified to obtain the coordinates of M, we have

$$ka\alpha + (ka\alpha - lc\gamma)t = a\alpha + (a\alpha - c\gamma)s$$

$$k\alpha + (k\alpha - l\gamma)t = \alpha + (\alpha - \gamma)s$$

Multiplying the second by a and subtracting, we conclude that $s = t$, so that

$$t = \frac{1-k}{k-l}, \qquad s = \frac{l(1-k)}{k-l}$$

By replacing a, α, s, t by b, β, u, v we obtain for the point L the parameter $u = v$ and, as before,

7.63 $$v = \frac{1-k}{k-l} = t, \qquad u = \frac{l(1-k)}{k-l} = s$$

The direction numbers of AB are $a\alpha - b\beta$, $\alpha - \beta$, while those of ML are

$$[a\alpha + (a\alpha - c\gamma)s] - [b\beta + (b\beta - c\gamma)u] = (a\alpha - b\beta)(1 + s)$$

and

$$[\alpha + (\alpha - \gamma)s] - [\beta + (\beta - \gamma)u] = (\alpha - \beta)(1 + s)$$

by 7.63; thus ML is parallel to AB as we wished to show.

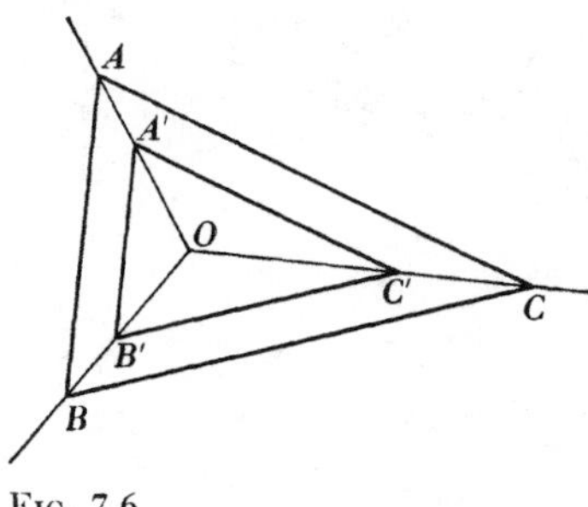

FIG. 7.6

Case (ii). If $A'B'$ is parallel to AB and $B'C'$ is parallel to BC, then it follows immediately that $A'C'$ is parallel to AC, though we cannot set $k = l$ in 7.63 to prove it.

Instead of introducing a different argument in each of the three cases of Desargues' theorem we could have used homogeneous coordinates. We illustrate the method in the following section.

EXERCISES

The simplest example of a non-Desarguesian plane is due to F. R. Moulton. With nonhomogeneous coordinates taken in the Euclidean plane π, a *modified line* is defined by the equation

$$x_2 = m(x_1 - a)f(x_2,m)$$

where the function f is defined as follows:

$$\begin{aligned} &\text{(i) if} \quad m \leqslant 0, \quad f(x_2,m) = 1 \\ &\text{(ii) if} \quad m > 0 \quad \text{and} \quad x_2 \leqslant 0, \quad f(x_2,m) = 1 \\ &\text{(iii) if} \quad m > 0 \quad \text{and} \quad x_2 > 0, \quad f(x_2,m) = \tfrac{1}{2} \end{aligned}$$

A modified line is identical with an ordinary line in π in the first case; in the other cases a modified line is made up of two "half-lines." Verify that:

1. Any two points P, Q uniquely determine a modified line PQ;
2. Two modified lines intersect in a unique point or are parallel;
3. Two modified lines are parallel if and only if the corresponding half-lines are parallel;
4. Desargues' theorem is *not* valid for all choices of the two triangles ABC and $A'B'C'$ in Moulton's geometry.

7.7 PAPPUS' THEOREM IN THE EUCLIDEAN PLANE

We have two special cases of Pappus' theorem in the Euclidean plane, but we shall consider first the general theorem 7.42 using homogeneous coordinates.

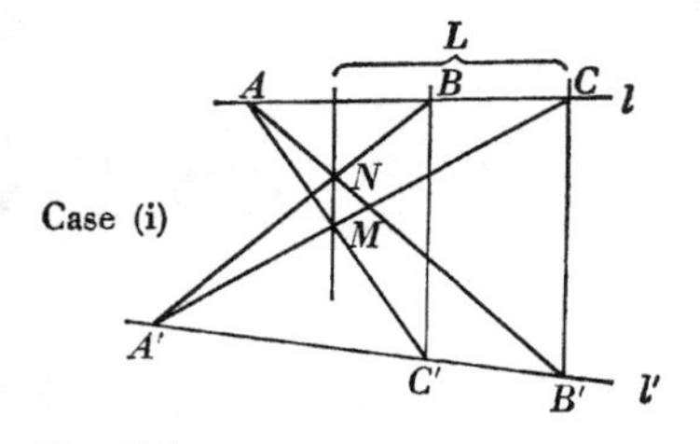

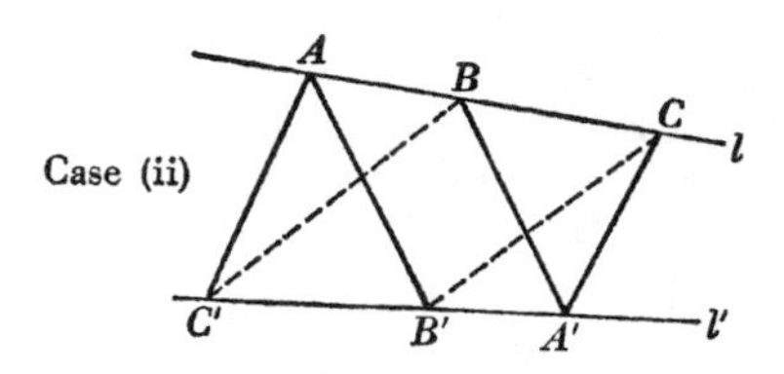

FIG. 7.7

If we take the origin at the intersection of l and l', we may take these lines as coordinate axes with equations $x_1 = 0$, $x_2 = 0$ respectively, so that the homogeneous coordinates of A, B, C may be taken to be $(0,a,1)$, $(0,b,1)$, $(0,c,1)$ and those of A', B', C' to be $(a',0,1)$, $(b',0,1)$, $(c',0,1)$

respectively. In such a choice of oblique axes we are utilizing the convenience of expression suggested in Section 5.1.

Setting up the necessary equations, e.g.,

$$\frac{x_1}{c'} + \frac{x_2}{b} = x_0 = \frac{x_1}{b'} + \frac{x_2}{c}$$

for the point L, we readily verify that the *homogeneous* coordinates of L, M, N may be taken to be

7.71 $$L \ : \ \frac{1}{b} - \frac{1}{c}, \ \frac{1}{b'} - \frac{1}{c'}, \ \frac{1}{bb'} - \frac{1}{cc'}$$

7.72 $$M \ : \ \frac{1}{c} - \frac{1}{a}, \ \frac{1}{c'} - \frac{1}{a'}, \ \frac{1}{cc'} - \frac{1}{aa'}$$

7.73 $$N \ : \ \frac{1}{a} - \frac{1}{b}, \ \frac{1}{a'} - \frac{1}{b'}, \ \frac{1}{aa'} - \frac{1}{bb'}$$

Since the third-order determinant Δ made up of these three sets of coordinates vanishes (the sum in each column being zero), we conclude that L, M, N are collinear and *this is independent of the vanishing or nonvanishing of the entries in the third column of* Δ (i.e., of the coordinate of homogeneity). Thus we have proved Pappus' theorem not only when all intersections L, M, N are finite, but also in cases (i) and (ii) of Figure 7.7.

EXERCISE

1. Set up the necessary equations to justify 7.71–7.73.

7.8 CROSS RATIO

There is one remarkable property of the correspondence established in Section 7.1 which again we prove only in the Euclidean plane. If the four lines OA, OB, OC, OD in Figure 7.8 have equations

7.81 $$x_2 = ax_1, \quad x_2 = bx_1, \quad x_2 = cx_1, \quad x_2 = dx_1$$

we may suppose that they are met by any transversal

7.82 $$x_2 = mx_1 + n$$

in the points A, B, C, D. If A_1, B_1, C_1, D_1 are the feet of the perpendiculars from A, B, C, D on Ox_1, then, from similar triangles,

7.83 $$\frac{AB \cdot CD}{CB \cdot AD} = \frac{A_1B_1 \cdot C_1D_1}{C_1B_1 \cdot A_1D_1}$$

where

$$A_1B_1 = \frac{n}{a - m} - \frac{n}{b - m} = \frac{n(b - a)}{(a - m)(b - m)}, \quad \text{etc.}$$

Assuming that $n \neq 0$, we conclude that the *cross ratio*

7.84 $$\{AC,BD\} = \frac{AB \cdot CD}{CB \cdot AD} = \frac{(b - a)(d - c)}{(b - c)(d - a)}$$

is independent of m, n and so of the choice of the transversal.

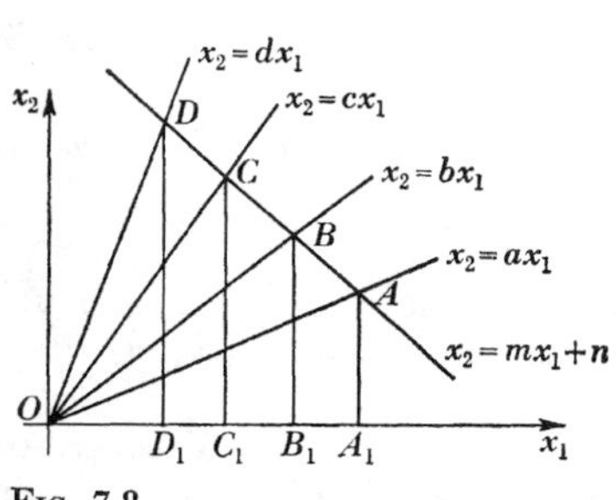

Fig. 7.8

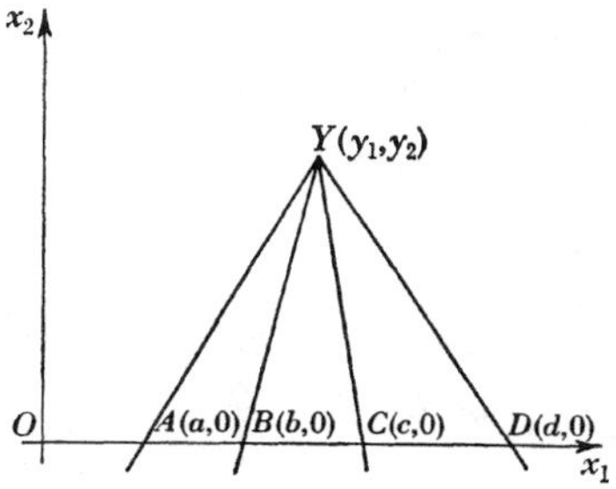

Fig. 7.9

But we can look at the problem slightly differently. Suppose now that the points $A(a,0)$, $B(b,0)$, $C(c,0)$, $D(d,0)$ are fixed and the vertex $Y(y_1,y_2)$ of the pencil of four lines is arbitrary (Figure 7.9). The slopes of the lines are easily seen to be

$$m_a = \frac{y_2}{y_1 - a}, \quad m_b = \frac{y_2}{y_1 - b}, \quad m_c = \frac{y_2}{y_1 - c}, \quad m_d = \frac{y_2}{y_1 - d}$$

and assuming that $y_2 \neq 0$, we have

7.85 $$\frac{(m_b - m_a)(m_d - m_c)}{(m_b - m_c)(m_d - m_a)} = \frac{(b - a)(d - c)}{(b - c)(d - a)}$$

Thus the same *cross ratio* is defined by 7.84 and 7.85 and this is independent not only of the choice of the transversal for a fixed pencil but also of the pencil for a fixed transversal. The cross ratio is a *projective invariant* and plays an important role in the further development of projective geometry.

In the particular case in which $\{AC,BD\} = -1$, the four points A, B, C, D are said to form an *harmonic range* and YA, YB, YC, YD an *harmonic pencil*. In this case

$$\{AC,BD\} = \{AC,BD\}^{-1} = \{AC,DB\}$$

and since we always have

$$\{AC,BD\} = \{BD,AC\} = \{CA,DB\}$$

the relationship is completely symmetrical with regard to the pairs A, C and B, D, which are called *harmonic conjugates* of each other.

EXERCISES

1. Prove that the internal and external bisectors of any angle form, with the arms of the angle, an harmonic pencil of lines meeting the base of the triangle in an harmonic range of points.

2. What can you say concerning the pencil of lines and range of points in Exercise 1 if the triangle in question is isosceles?

3. Using homogeneous coordinates, verify that 7.84 may be written in the form

$$\{XY,ZT\} = \frac{(z_1 - x_1)(t_1 - y_1)}{(z_1 - y_1)(t_1 - x_1)} = \frac{(z_1x_0 - x_1z_0)(t_1y_0 - y_1t_0)}{(z_1y_0 - y_1z_0)(t_1x_0 - x_1t_0)}$$

4. Prove that $\{XY,ZT\}$ is harmonic for any points $X(x_1,1)$, $Y(y_1,1)$, $Z = [\frac{1}{2}(x_1 + y_1), 1]$, $T = (1,0)$, and show that the point Z is uniquely defined by X, Y.

5. If the midpoint P' of the segment AB in Figure 7.4 is defined by the condition that $\{AB,P'P_\infty\} = -1$, and the midpoint Q of BD by the condition that $\{BD,Q'Q_\infty\} = -1$, prove that $P'Q_\infty$ and $Q'P_\infty$ meet the opposite sides of the parallelogram $ABCD$ in their midpoints.

6. If A' is the midpoint of BC, B' the midpoint of CA, and C' the midpoint of AB in the triangle ABC of Figure 10, prove that $C'B' \| BC$ and also that $BA' = C'B' = A'C$ for all l_∞ not passing through A, B, C.

   Solution. Assume that $\{BC,A'A_\infty\} = -1$, so that projecting from C_∞ we have $\{AC,B'B_\infty\} = -1$ and from A_∞ we have $\{AB,C'C_\infty\} = -1 = \{BA,C'C_\infty\}$. It follows that $C'A'$ must pass through B_∞, so that $BA' = C'B' = A'C$.

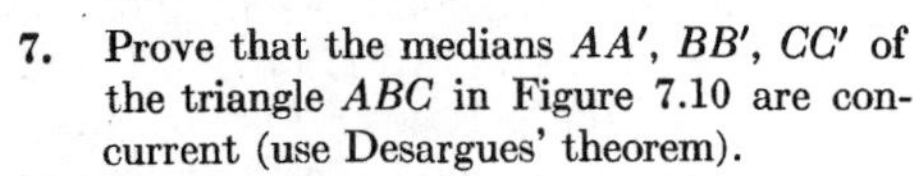

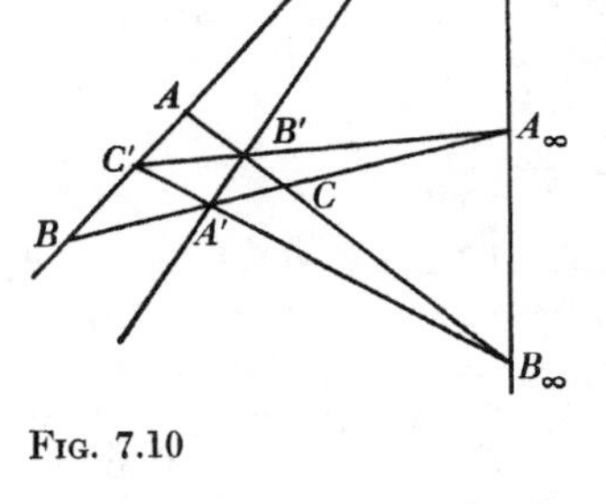

FIG. 7.10

7. Prove that the medians AA', BB', CC' of the triangle ABC in Figure 7.10 are concurrent (use Desargues' theorem).

8. By actual substitution, prove that the cross ratio of four points is invariant under any linear transformation of the coordinates.

8 Geometry on the Sphere

8.1 SPHERICAL TRIGONOMETRY

Though we live on the surface of the earth all our days and make maps which guide our automobiles and our airplanes at ever increasing speed, yet most people are not clear as to the relationship between plane and spherical geometry. Undoubtedly the explanation of this lies in the fact that our geometrical ideas stem from Euclid and the concept of a flat world was acceptable long after his time. To understand this mapping process and to introduce the study of a non-Euclidean metric, we examine first *spherical trigonometry* which is the basis of all large-scale surveying.

Euclid's definition of a "straight" line as the shortest distance between two points introduces the idea of a *metric* into geometry. When we think of a surface Σ embedded in 3-dimensional Euclidean space, this notion of the "shortest distance" between two points on Σ leads to a unique curve called a *geodesic* on Σ. If a thread be stretched between two points P and Q on a *sphere* Σ, then this thread will lie along the great circle joining P and Q, and this is a geodesic on the sphere.

For simplicity, let us assume that our sphere Σ has its center at the origin O with unit radius, and so has equation

8.11 $$x_1^2 + x_2^2 + x_3^2 = 1$$

If A, B, C are any three points on Σ, then we call the intersections of the planes OAB, OAC, OBC with Σ a *spherical triangle*, as in the accompanying figure. The metric we adopt on Σ is that of the Euclidean space in which Σ is embedded, so that the "length" of the side AB is determined by the angle $AOB = c$. In fact, these angles a, b, c, which are subtended by the sides BC, CA, AB at O, yield precisely the desired lengths if they are expressed in *radians* (i.e., as fractions of 2π). We define the *angle* A of the spherical triangle ABC to be that between the tangents AD and AE to the great circles AB and AC. We remove the ambiguity as to which of the two possible triangles we are referring to by the convention that *every angle of the triangle ABC shall be less than* π.

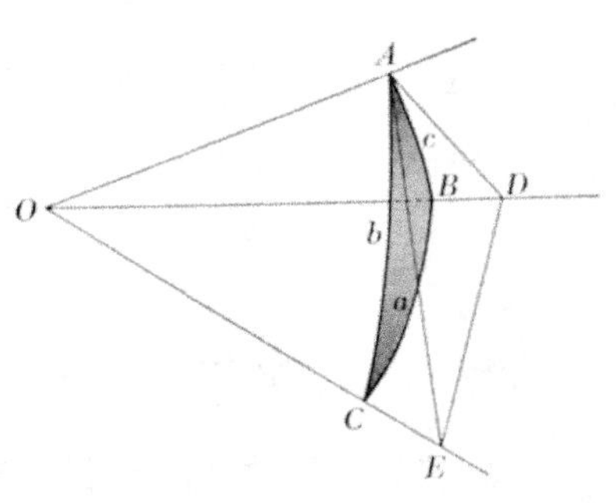

FIG. 8.1

With these definitions it follows from Pythagoras' theorem applied to the triangles ADE and ODE that

$$AD^2 + AE^2 - 2AD \cdot AE \cos A = OD^2 + OE^2 - 2OD \cdot OE \cos a$$

so that

$$OD \cdot OE \cos a = OA^2 + AD \cdot AE \cos A$$

and

$$\cos a = \frac{OA}{OE} \cdot \frac{OA}{OD} + \frac{AE}{OE} \cdot \frac{AD}{OD} \cos A$$

Referring to Figure 8.1 we have our basic formula:

8.12 $$\cos a = \cos b \cos c + \sin b \sin c \cos A$$

This, and a host of more complicated relations (cf. exercises of Section 8.2) were derived in the sixteenth and seventeenth centuries with the expansion of navigation and the need to fix latitude and longitude on long voyages of exploration. The following consequence of 8.12 resembles a familiar formula of plane trigonometry:

8.13 $$\frac{\sin A}{\sin a} = \frac{\sin B}{\sin b} = \frac{\sin C}{\sin c} = \frac{2\{\sin s \sin (s-a) \sin (s-b) \sin (s-c)\}^{1/2}}{\sin a \sin b \sin c}$$

where $2s = a + b + c$.

Proof. From 8.12 we have*

$$1 - \cos A = 1 - \frac{\cos a - \cos b \cos c}{\sin b \sin c} = \frac{\cos (b-c) - \cos a}{\sin b \sin c}$$

so that

$$\sin^2 \frac{A}{2} = \frac{\sin \frac{1}{2}(a+b-c) \sin \frac{1}{2}(a-b+c)}{\sin b \sin c}$$

Similarly,

$$\cos^2 \frac{A}{2} = \frac{\sin \frac{1}{2}(a+b+c) \sin \frac{1}{2}(b+c-a)}{\sin b \sin c}$$

Thus

$$\sin A = 2 \sin \frac{A}{2} \cos \frac{A}{2}$$

$$= \frac{2\{\sin s \sin (s-a) \sin (s-b) \sin (s-c)\}^{1/2}}{\sin b \sin c}$$

which yields 8.13, since the expression for $(\sin A/\sin a)$ is symmetrical in each of a, b, c.

It is important to observe that we can have a spherical triangle in which $A = B = C = \pi/2$, so that *the sum of the angles is in this case* $> \pi$. We shall see shortly that this striking difference from plane geometry holds for *every* spherical triangle.

8.2 THE POLAR TRIANGLE

So far, we have not drawn attention to the fact that two points B, C define not one but two (complementary) great circular arcs on Σ. Associated with the great circle BC in Figure 8.2 are the two *poles* A', A'' which are the

* The basic formulas of elementary trigonometry which we require are as follows:

$\sin (\alpha + \beta) = \sin \alpha \cos \beta + \cos \alpha \sin \beta$; $\quad \sin (\alpha - \beta) = \sin \alpha \cos \beta - \cos \alpha \sin \beta$

$\cos (\alpha + \beta) = \cos \alpha \cos \beta - \sin \alpha \sin \beta$; $\quad \cos (\alpha - \beta) = \cos \alpha \cos \beta + \sin \alpha \sin \beta$

from which we deduce that

$$\sin 2\alpha = 2 \sin \alpha \cos \alpha$$

$$\cos 2\alpha = \cos^2 \alpha - \sin^2 \alpha = 2 \cos^2 \alpha - 1 = 1 - 2 \sin^2 \alpha$$

and also that

$$\cos \alpha - \cos \beta = -2 \sin \frac{\alpha + \beta}{2} \sin \frac{\alpha - \beta}{2}$$

intersections with Σ of the diameter through O such that $A'OA''$ is normal to the plane OBC. Corresponding to the other two sides of the spherical triangle we have the poles B', B'' and C', C''.

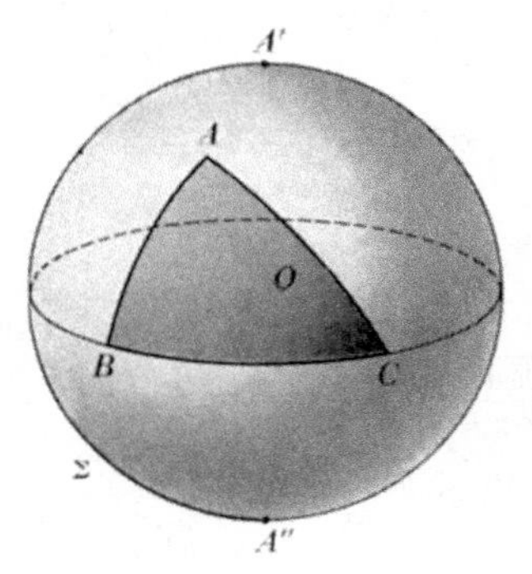

FIG. 8.2

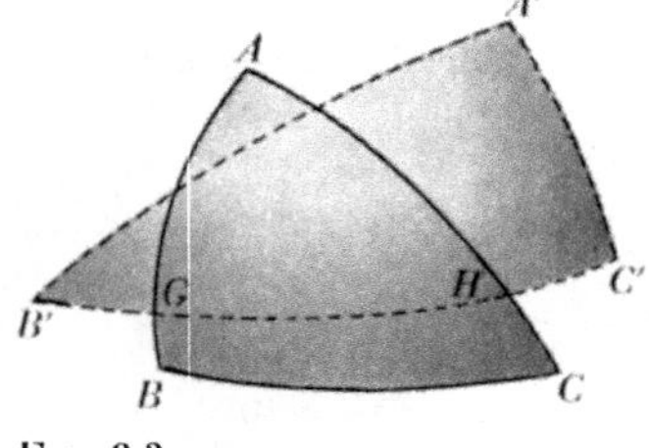

FIG. 8.3

It is clear that either A' or A'' will lie in the hemisphere which contains A, defined by the great circle BC; let us suppose it to be A'. Similarly, we distinguish B' and C' and we call $A'B'C'$ the *polar triangle* of ABC. If we examine Figure 8.3 it appears that AB', AC'; BA', BC'; CA', CB' are all quadrants of great circles, so that *ABC is the polar triangle of $A'B'C'$.* Moreover, GH is precisely the measure of the angle between the planes OAB and OAC which we have denoted by A, while $B'H = C'G = \pi/2$. Thus

$$B'C' = a' = \pi - A, \quad C'A' = b' = \pi - B, \quad A'B' = c' = \pi - C$$

From the similar properties of ABC we conclude that

$$A' = \pi - a, \quad B' = \pi - b, \quad C' = \pi - c$$

If now we substitute these expressions in the analogue of 8.12 for the polar triangle we have

$$\cos(\pi - A) = \cos(\pi - B)\cos(\pi - C) + \sin(\pi - B)\sin(\pi - C)\cos(\pi - a)$$

or

8.21 $$\cos A = -\cos B \cos C + \sin B \sin C \cos a$$

which we can think of as the *dual* of 8.12.

There is another approach to spherical trigonometry which depends on properties of the vector product. If we write $\overrightarrow{OA} = \boldsymbol{A}$, $\overrightarrow{OB} = \boldsymbol{B}$, $\overrightarrow{OC} = \boldsymbol{C}$, then the Lagrange identity (cf. Exercise 5, Section 5.3) leads to the conclusion

$$\text{8.22} \qquad (\boldsymbol{A}\times\boldsymbol{B})\cdot(\boldsymbol{A}\times\boldsymbol{C}) = \begin{vmatrix} \boldsymbol{A}\cdot\boldsymbol{A} & \boldsymbol{A}\cdot\boldsymbol{C} \\ \boldsymbol{B}\cdot\boldsymbol{A} & \boldsymbol{B}\cdot\boldsymbol{C}\end{vmatrix} = \begin{vmatrix} 1 & \cos b \\ \cos c & \cos a\end{vmatrix}$$

$$= \cos a - \cos b \cos c$$

since $|\boldsymbol{A}| = |\boldsymbol{B}| = |\boldsymbol{C}| = 1$. But we also have

$$\text{8.23} \qquad (\boldsymbol{A}\times\boldsymbol{B})\cdot(\boldsymbol{A}\times\boldsymbol{C}) = |\boldsymbol{A}\times\boldsymbol{B}||\boldsymbol{A}\times\boldsymbol{C}| \cos A$$

$$= \sin c \sin b \cos A$$

by 1.45 and 5.372. Taken together, 8.22 and 8.23 yield 8.12.

In order to derive 8.13, we use 5.372 again to yield

$$\text{8.24} \qquad |(\boldsymbol{A}\times\boldsymbol{B})\times(\boldsymbol{A}\times\boldsymbol{C})| = |\boldsymbol{A}\times\boldsymbol{B}||\boldsymbol{A}\times\boldsymbol{C}| \sin A$$

$$= \sin c \sin b \sin A$$

By evaluating $|(\boldsymbol{A}\times\boldsymbol{B})\times(\boldsymbol{A}\times\boldsymbol{C})|$ differently, we obtain a quantity σ which is unchanged by permutation of the vectors $\boldsymbol{A}$, $\boldsymbol{B}$, $\boldsymbol{C}$ in any manner. Thus we conclude that

$$\text{8.25} \qquad \frac{\sin A}{\sin a} = \frac{\sin B}{\sin b} = \frac{\sin C}{\sin c} = \frac{\sigma}{\sin a \sin b \sin c}$$

where σ can be shown to have the value indicated in 8.13.

EXERCISES

1. Derive the special cases of 8.12, 8.13, and 8.21 for a spherical triangle in which $A = \pi/2$.
2. Prove that

$$\sin^2 \frac{a}{2} = \frac{-\cos \frac{1}{2}(A+B+C) \cos \frac{1}{2}(B+C-A)}{\sin B \sin C} = \frac{-\cos S \cos (S-A)}{\sin B \sin C}$$

where $2S = A + B + C$.

3. Prove that

$$\cos \frac{a}{2} = \left\{\frac{\cos (S-B) \cos (S-C)}{\sin B \sin C}\right\}^{1/2}$$

and thence that

$$\sin a = \frac{2\{-\cos S \cos (S-A) \cos (S-B) \cos (S-C)\}^{1/2}}{\sin B \sin C}$$

4. Show that

$$\tan \frac{A}{2} \tan \frac{B}{2} = \frac{\sin (s-c)}{\sin s}$$

5. Verify the correctness of Napier's *Analogies*

$$\tan \tfrac{1}{2}(A+B) \tan \tfrac{1}{2}C = \frac{\cos \frac{1}{2}(a-b)}{\cos \frac{1}{2}(a+b)}$$

$$\tan \tfrac{1}{2}(A-B) \tan \tfrac{1}{2}C = \frac{\sin \frac{1}{2}(a-b)}{\sin \frac{1}{2}(a+b)}$$

by expanding and substituting from Exercise 4. By applying these formulas to the polar triangle, obtain their duals.

6. Describe a procedure for solving a spherical triangle in which the given elements are (i) a, b, c; (ii) A, B, C; (iii) a, b, C; (iv) A, B, a. Can you arrange that each problem be solved conveniently by using logarithms?

8.3 AREA OF A SPHERICAL TRIANGLE

Consider now Figure 8.4, in which is represented the hemisphere containing A and defined by the great circle BC. If we suppose the sphere Σ to have radius r and area $4\pi r^2$, then the area of the *lune* made up of the parts ABC and AB_1C_1 is $(A/\pi)2\pi r^2$. Similarly, the area of $ABC + AB_1C$ is $(B/\pi)2\pi r^2$, and the area of $ABC + ABC_1$ is $(C/\pi)2\pi r^2$.

Thus,

8.31
$$\text{Area of triangle } ABC = \frac{1}{2}\left(\frac{A}{\pi} + \frac{B}{\pi} + \frac{C}{\pi} - 1\right)2\pi r^2$$
$$= (A + B + C - \pi)r^2$$

which is a famous result that leads to the designation of the quantity $A + B + C - \pi$ as the *spherical excess.*

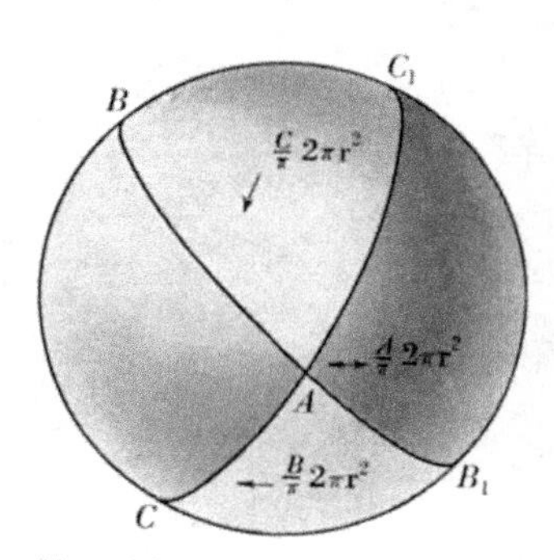

FIG. 8.4

Here we have a sharp distinction between geometry on the surface of a sphere and in the Euclidean plane. In the first place the area of a hemisphere is finite ($2\pi r^2$), while that of the Euclidean plane is not. If we think of r as tending to infinity and the area of the triangle ABC remaining constant, then it is clear that $A + B + C - \pi$ must tend to zero, which yields the Euclidean theorem concerning the sum of the angles of a triangle.

In the second place, any two great circles will intersect (once in any given hemisphere) so that there is no such thing as *parallelism* on the sphere.

Finally, given any two great circles on the sphere there exists a uniquely determined great circle through their poles which is orthogonal to each great circle. The analogous statement concerning lines in the plane is true only if the lines are parallel, and then there is an infinite number of common perpendiculars (see the end of Section 8.8).

EXERCISES

1. Prove that no two triangles on a sphere Σ have equal angles, i.e., are "similar," unless their sides are equal also.

2. From the property that the medians of a triangle are concurrent in Euclidean geometry deduce the corresponding property of a triangle on the surface of a sphere.

3. If $A = B = C = \pi/2$, express the area of the spherical triangle ABC as a fraction of the total area of the sphere Σ.

4. Show that the two restrictions on a spherical triangle ABC, (i) that every angle be less than π, (ii) that every side be less than πr, are equivalent.

8.4 THE INVERSION TRANSFORMATION

Though we have seen how geometry on the sphere can approximate that in the Euclidean plane for r sufficiently large, yet we would like to establish a closer correspondence between the two. To this end we investigate the quadratic transformation known as *inversion*.

In order to define the transformation we consider it first with reference to the circle $\mathcal{C}_0$,

$$x_1^2 + x_2^2 = a^2$$

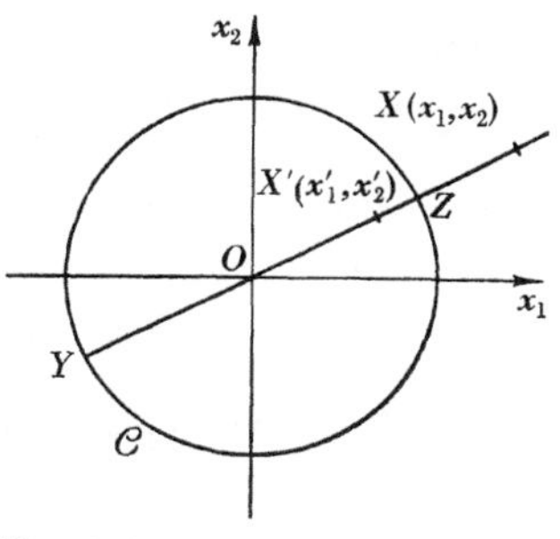

FIG. 8.5

If $X(x_1,x_2)$ is any point different from O, then OX will meet $\mathcal{C}_0$ in Y, Z, and the *inverse point* X' *of* X *with regard to* $\mathcal{C}_0$ is defined to be the harmonic conjugate of X with regard to Y and Z. Thus

8.41
$$\{YZ,XX'\} = \frac{YX \cdot ZX'}{YX' \cdot ZX} = \frac{(YO + OX)(ZO + OX')}{(YO + OX')(ZO + OX)} = -1$$

which on expanding and simplifying leads to the alternative definition

8.42
$$OX \cdot OX' = \sqrt{x_1^2 + x_2^2}\,\sqrt{x_1'^2 + x_2'^2} = a^2$$

Rewriting the equation 8.41 in the form

$$\frac{XY \cdot X'Z}{XZ \cdot X'Y} = \frac{XY(XZ - XX')}{XZ(XY - XX')} = -1$$

we have

$$\text{8.43} \qquad \frac{2}{XX'} = \frac{1}{XY} + \frac{1}{XZ}, \qquad \frac{2}{X'X} = \frac{1}{X'Y} + \frac{1}{X'Z}$$

If we substitute from the relation

$$\frac{x_1}{x_2} = \frac{x_1'}{x_2'}$$

in 8.42, we obtain the equations

$$\text{8.441} \qquad x_1' = \frac{x_1 a^2}{x_1^2 + x_2^2}, \qquad x_2' = \frac{x_2 a^2}{x_1^2 + x_2^2}$$

of the *quadratic* transformation known as *inversion*, with the inverse transformation given by the equations

$$\text{8.442} \qquad x_1 = \frac{x_1' a^2}{x_1'^2 + x_2'^2}, \qquad x_2 = \frac{x_2' a^2}{x_1'^2 + x_2'^2}$$

Leaving the geometrical properties of the transformation to be developed in the next section, let us consider the following

Problem. Find the locus of the harmonic conjugate X of a fixed point X' with regard to the circle $\mathcal{C}$, $x_1^2 + x_2^2 = a^2$.

Of course the significance of the problem is that we take an *arbitrary* line through X' meeting $\mathcal{C}$ in Y and Z and look for the locus of the harmonic conjugate X of X'. Such a line could be written in parametric form:

$$\text{8.45} \qquad x_1 = x_1' + l_1 t, \qquad x_2 = x_2' + l_2 t$$

where t is the length of the segment $X'X$ and l_1, l_2 are the direction cosines of the line. Substituting the expressions 8.45 in the equation of $\mathcal{C}$ we have

$$\text{8.46} \qquad (x_1'^2 + x_2'^2 - a^2)\frac{1}{t^2} + \frac{2}{t}(x_1' l_1 + x_2' l_2) + 1 = 0$$

If now we require that the roots $1/t$ shall satisfy 8.43, we have

$$\frac{1}{t} = -\frac{x_1' l_1 + x_2' l_2}{x_1'^2 + x_2'^2 - a^2}$$

so that after cross-multiplying and substituting from 8.45 we have the required locus,

$$\text{8.47} \qquad x_1 x_1' + x_2 x_2' = a^2$$

This line is perpendicular to OX' and is called the *polar* (*line*) of X' with regard to the circle $\mathcal{C}$; X' is the *pole* of 8.47.

EXERCISES

1. Prove that if the polar of X' with regard to a circle $\mathcal{C}$ passes through Y', then the polar of Y' passes through X'.

2. By a method similar to that used above, find the polar of X' with regard to each of the three conics

$$\frac{x_1^2}{a^2} + \frac{x_2^2}{b^2} = 1, \quad \frac{x_1^2}{a^2} - \frac{x_2^2}{b^2} = 1, \quad x_2^2 = 4px_1$$

3. Prove that the polar line of the focus of a conic is the corresponding directrix.
4. If the polar of L with regard to a given conic $\mathcal{C}$ is l and the polar of M with regard to $\mathcal{C}$ is m, prove that the pole of LM is the point of intersection of l and m.
5. Using homogeneous coordinates, find the polar lines of the points at infinity (1,0,0), (0,1,0) with regard to the ellipse

$$\frac{x_1^2}{a^2} + \frac{x_2^2}{b^2} = x_0^2$$

In what point do these polar lines intersect?

6. Use the result of Exercise 5 to define the center of a conic in affine geometry.

8.5 GEOMETRICAL PROPERTIES OF INVERSION

Since the equation of an arbitrary circle $\mathcal{C}$ can be written in the form

8.51
$$x_1^2 + x_2^2 + 2gx_1 + 2fx_2 + c = 0$$

we obtain the *inverse curve* $\mathcal{C}'$ with regard to the circle $\mathcal{C}_0$ by substituting from 8.442 and multiplying by $x_1'^2 + x_2'^2 \neq 0$:

8.52
$$a^4 + 2ga^2x_1' + 2fa^2x_2' + c(x_1'^2 + x_2'^2) = 0$$

Thus,

8.53 *The inverse of a circle $\mathcal{C}$ is a circle $\mathcal{C}'$, unless $\mathcal{C}$ passes through the origin, in which case the inverse of $\mathcal{C}$ is a line. Conversely, the inverse of a line l is a circle $\mathcal{C}$ through O, unless l passes through O when l is mapped upon itself by the transformation.*

More generally, we can invert with regard to any circle of radius a, the center of which is called the *center of inversion* and a the *radius of inversion.* Clearly, any point on the circle of inversion remains fixed under the transformation.

The three-dimensional analogue of inversion in a circle is inversion in a *sphere* Σ_0,

$$x_1^2 + x_2^2 + x_3^2 = a^2$$

and the analogues of 8.441 and 8.442 are easily seen to be

8.541
$$x_1' = \frac{x_1a^2}{x_1^2 + x_2^2 + x_3^2}, \qquad x_2' = \frac{x_2a^2}{x_1^2 + x_2^2 + x_3^2}$$
$$x_3' = \frac{x_3a^2}{x_1^2 + x_2^2 + x_3^2}$$

and

8.542
$$x_1 = \frac{x_1'a^2}{x_1'^2 + x_2'^2 + x_3'^2}, \qquad x_2 = \frac{x_2'a^2}{x_1'^2 + x_2'^2 + x_3'^2}$$
$$x_3 = \frac{x_3'a^2}{x_1'^2 + x_2'^2 + x_3'^2}$$

Note that points on the circle or sphere of inversion remain *fixed*, and these are the only fixed points of the transformation.

As in the case of a circle, the sphere with equation

8.55
$$x_1^2 + x_2^2 + x_3^2 + 2gx_1 + 2fx_2 + 2hx_3 + c = 0$$

inverts into the sphere with equation

8.56
$$a^4 + 2ga^2x_1' + 2fa^2x_2' + 2ha^2x_3' + c(x_1'^2 + x_2'^2 + x_3'^2) = 0$$

under the transformations 8.542. Thus,

8.57 *The inverse of a sphere Σ is a sphere Σ', unless Σ passes through the center of inversion O, in which case the inverse is a plane. Conversely, the inverse of a plane π is a sphere Σ passing through O unless π itself passes through O, in which case π is mapped upon itself by the transformation.*

The following special case of 8.53 and 8.57 is of some interest.

8.58 *Any circle (sphere) orthogonal to the circle (sphere) of inversion is its own inverse.*

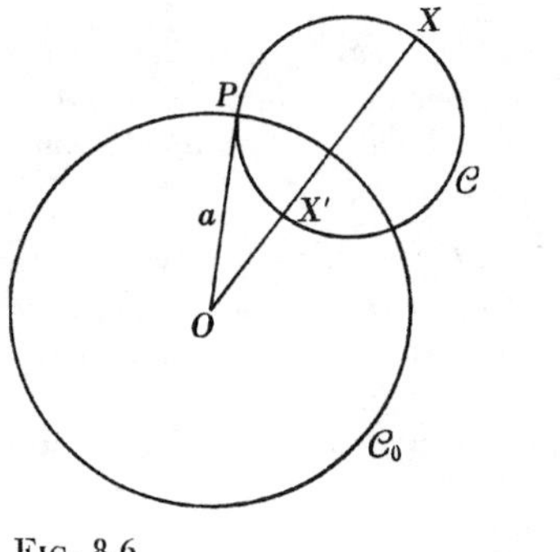

FIG. 8.6

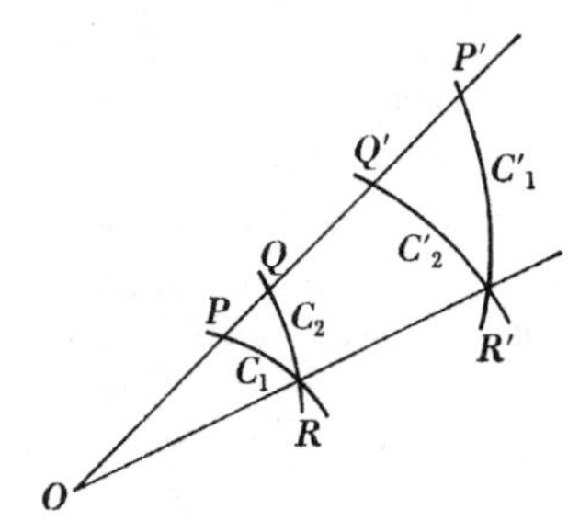

FIG. 8.7

Proof. The proof is an immediate consequence of the fact that if $\mathcal{C}$ is orthogonal to $\mathcal{C}_0$, then OP is tangent to $\mathcal{C}$ and

$$OP^2 = OX \cdot OX' = a^2$$

so that the points X, X' on $\mathcal{C}$, collinear with O, are inverse points with regard to $\mathcal{C}_0$. The argument for spheres is the same.

There is one further property which is important:

8.59 *Inversion is a conformal transformation.*

This means that if two curves C_1 and C_2 intersect in R at an angle α, then their inverses C_1' and C_2' will also intersect in R', the inverse of R, and at the same angle α.

Proof. By definition,

$$OP \cdot OP' = OQ \cdot OQ' = OR \cdot OR'$$

so that

$$\frac{OP}{OR} = \frac{OR'}{OP'} \quad \text{and} \quad \frac{OQ}{OR} = \frac{OR'}{OQ'}$$

From similar triangles, it follows that

$$\angle OPR = \angle OR'P', \qquad \angle OQR = \angle OR'Q'$$

so that

$$\angle PRQ = \angle OPR - \angle OQR = \angle OR'P' - \angle OR'Q' = \angle P'R'Q'$$

Since the angle between the curves C_1, C_2 is defined to be the limit α of $\angle PRQ$, as $P \to R$, it follows that the limit of $\angle P'R'Q'$ is also α, proving the theorem.

EXERCISES

1. Invert the property of circles which states that the angle in a semicircle is a right angle.
2. Show that:

   (a) Inverse points with regard to a circle $\mathcal{C}$ invert into inverse points with regard to $\mathcal{C}'$.

   (b) The limit points of a coaxal system are inverse points with regard to every circle of the system.

   (c) A system of nonintersecting coaxal circles may be inverted into a system of concentric circles.

   (d) A system of intersecting coaxal circles can be inverted into a system of concurrent lines.

8.6 STEREOGRAPHIC PROJECTION

If we designate by N and S the "north" and "south" poles of a sphere Σ, the process known as *stereographic projection* consists of

(i) projecting Σ onto the tangent plane at N from the point S, or
(ii) projecting Σ onto the equatorial plane from the point N.

We consider first case (i), and show that the effect of projection is identical with that of inverting Σ with respect to a sphere Σ_0 with center S, passing through N. Clearly the plane π tangent to Σ at N is also the plane tangent to Σ_0 at N, so that the inverse of Σ is π. Moreover, any plane through NS meets Σ in a line of "longitude" so that such lines invert into lines through N in π. Finally, any small circle on Σ is the intersection of a plane ω with Σ, so that such small circles invert into the intersection of a sphere with π, i.e., into a circle in π. Thus lines of latitude on Σ project into concentric circles in π. Since lines of latitude and longitude intersect orthogonally on Σ, the inverse curves will also have this property, and we can describe the position of a point P' on the map

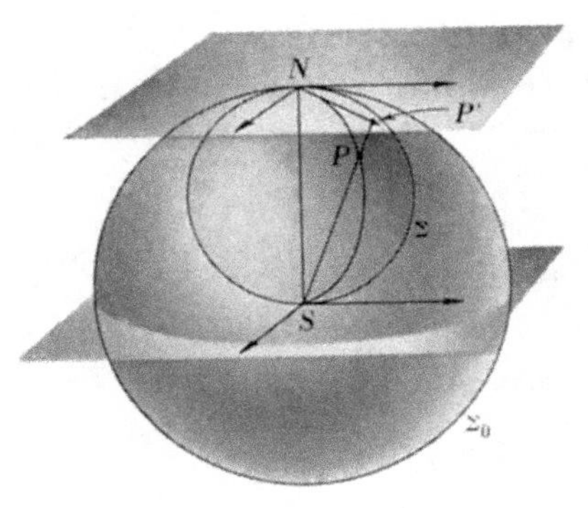

FIG. 8.8

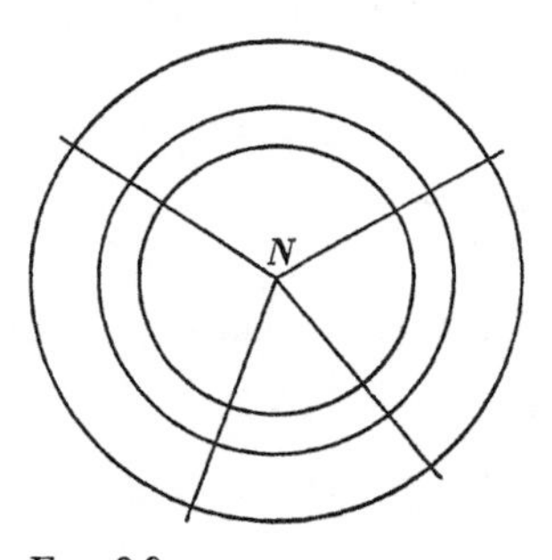

FIG. 8.9

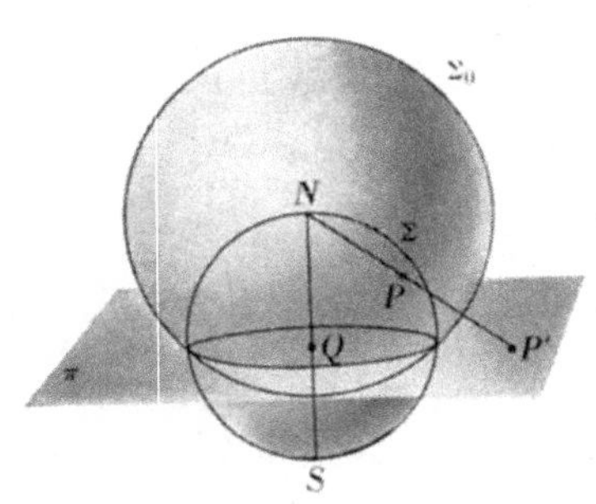

FIG. 8.10

in the same way as P was described on the surface of Σ. There is an increasing amount of distortion, however, as we approach the circumference of the map, since the Euclidean metric in π is different from that on the surface of Σ.

In case (ii) a similar relation between projection and inversion holds, except that the sphere of inversion Σ_0 has its center at N and passes through the equator of Σ. Great circles on Σ through S project into lines through Q, the center of Σ in π, while small circles on Σ project into circles in π. Conformality is preserved in both cases by the appropriate generalization of 8.59, and distortion again increases as we move away from the point Q.

8.7 ELLIPTIC GEOMETRY

That only certain great circles on the sphere project into lines in π is an undesirable feature of stereographic projection which must be tolerated to obtain conformality. But from an abstract geometrical point of view conformality is unnecessary, and by projection from the center of the sphere onto a tangent plane every great circle will project into a line.

Let us proceed analytically, taking the equation of Σ to be

$$x_1^2 + x_2^2 + x_3^2 = r^2$$

The equation of the plane π tangent to Σ at $\overline{O}$, in Figure 8.11, may be taken to be

$$x_3 = r$$

and we choose $\overline{O}\overline{x}_1$, $\overline{O}\overline{x}_2$ parallel to Ox_1, Ox_2. We are in exactly the position we considered in such detail in the preceding chapter: the coordinates

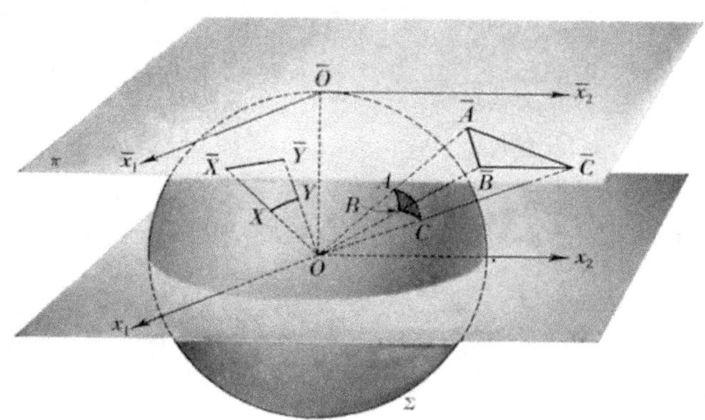

FIG. 8.11

(x_1,x_2,x_3) of any point X on Σ are related to those of its projection $\overline{X}$ in the tangent plane π by the relations

8.71
$$\frac{\overline{x}_1}{x_1} = \frac{\overline{x}_2}{x_2} = \frac{r}{x_3}$$

But here we have a *metric* which derives from that on the surface of the sphere Σ and which we can transfer in a nonambiguous manner to the tangent plane.

Using the notion of the inner product in 1.45, let us define the distance between two points X, Y on Σ to be $d(XY)$ so that

8.72 $$d(XY) = r\cos^{-1}\left\{\frac{x_1y_1 + x_2y_2 + x_3y_3}{\sqrt{x_1^2 + x_2^2 + x_3^2}\sqrt{y_1^2 + y_2^2 + y_3^2}}\right\}$$

8.73 $$= r\cos^{-1}\left\{\frac{\bar{x}_1\bar{y}_1 + \bar{x}_2\bar{y}_2 + r^2}{\sqrt{\bar{x}_1^2 + \bar{x}_2^2 + r^2}\sqrt{\bar{y}_1^2 + \bar{y}_2^2 + r^2}}\right\}$$

taken positive. If, now, we denote the *distance* between $\bar{X}$ and $\bar{Y}$ in π by $d(\bar{X}\bar{Y})$, this is determined by setting

8.74 $$d(\bar{X}\bar{Y}) = d(XY)$$

It follows immediately from 8.73 that $d(\bar{X}\bar{Y})$ *is always finite* for $\bar{X}$ fixed and any $\bar{Y}$, and *every pair of lines intersects.* We complete our definition of the metric in the tangent plane by defining the measure of an angle to be the same as that on the sphere Σ, so that

8.75 $$\Delta\bar{A}\bar{B}\bar{C} = \Delta ABC = (A + B + C - \pi)r^2 > 0$$

Thus all the formulas of Sections 8.1–8.3 applicable to the spherical triangle ABC apply also to $\bar{A}\bar{B}\bar{C}$, and the resulting geometry in π is called *elliptic.*

Actually, our mapping of the sphere Σ on the tangent plane is 2:1, since two diametrically opposite points on the sphere are mapped on the same point of the plane. Thus *the area of the elliptic plane is finite and equal to* $2\pi r^2$. It follows immediately from 8.75 that

8.76 *In any triangle ABC in elliptic geometry,*

$$A + B + C > \pi$$

In order to study analytically the effect of allowing r to increase, we write 8.73 in the form

8.77 $$d(\bar{X}\bar{Y}) = r\sin^{-1}\frac{1}{r}\left\{\frac{r^{-2}(\bar{x}_1\bar{y}_2 - \bar{x}_2\bar{y}_1)^2 + (\bar{x}_1 - \bar{y}_1)^2 + (\bar{x}_2 - \bar{y}_2)^2}{[r^{-2}(\bar{x}_1^2 + \bar{x}_2^2) + 1][r^{-2}(\bar{y}_1^2 + \bar{y}_2^2) + 1]}\right\}^{1/2}$$

taken positive.

Since $\lim_{r\to 0}\frac{1}{x}\sin^{-1}x = 1$, we conclude that

$$\lim_{r\to\infty} d(\bar{X}\bar{Y}) = \sqrt{(\bar{x}_1 - \bar{y}_1)^2 + (\bar{x}_2 - \bar{y}_2)^2}$$

which is Pythagoras' theorem, and *Euclidean geometry is a limiting case of elliptic or spherical geometry.*

8.8 HYPERBOLIC GEOMETRY

Suppose we replace r in 8.75 by ir, where $i^2 = -1$; then the condition 8.76 becomes

8.81 *In any triangle ABC in hyperbolic geometry*

$$A + B + C < \pi$$

Moreover, if we wish our projection to be real we must also replace x_3 by ix_3 in 8.71 so that the sphere becomes an hyperboloid of two sheets with equation

8.821 $$x_1^2 + x_2^2 - x_3^2 = -r^2$$

The enveloping or asymptotic cone of this hyperboloid with equation

8.822 $$x_1^2 + x_2^2 - x_3^2 = 0$$

cuts the tangent plane π: $x_3 = r$ in the circle

8.823 $$\mathcal{A}: \ \bar{x}_1^2 + \bar{x}_2^2 = r^2$$

called the *absolute* in π. By such a substitution the distance function 8.77 becomes

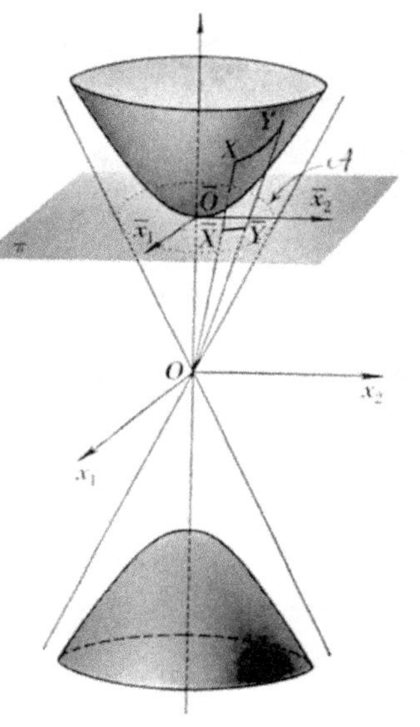

FIG. 8.12

8.83 $$d(\overline{XY}) = ir \sin^{-1} \frac{1}{ir} \left\{ \frac{-r^{-2}(\bar{x}_1\bar{y}_2 - \bar{x}_2\bar{y}_1) + (\bar{x}_1 - \bar{y}_1)^2 + (\bar{x}_2 - \bar{y}_2)^2}{[-r^{-2}(\bar{x}_1^2 + \bar{x}_2^2) + 1][-r^{-2}(\bar{y}_1^2 + \bar{y}_2^2) + 1]} \right\}^{1/2}$$

taken positive.

In order to see the further consequences of the change of r into ir we must examine how it affects the *spherical* functions

$$\cos\theta = \frac{1}{2}(e^{i\theta} + e^{-i\theta}), \qquad \sin\theta = \frac{1}{2i}(e^{i\theta} - e^{-i\theta})$$

If we define the corresponding *hyperbolic* functions by the formulas

$$\cosh\theta = \tfrac{1}{2}(e^{\theta} + e^{-\theta}), \qquad \sinh\theta = \tfrac{1}{2}(e^{\theta} - e^{-\theta})$$

then it is easy to see that $i \sin\theta = \sinh i\theta$ so that

$$i \sin^{-1}\frac{x}{i} = \sinh^{-1} x$$

and we can rewrite 8.83 in the real form

$$d(\bar{X}\bar{Y}) = r \sinh^{-1} \frac{1}{r} \left\{ \frac{-r^{-2}(\bar{x}_1\bar{y}_2 - \bar{x}_2\bar{y}_1) + (\bar{x}_1 - \bar{y}_1)^2 + (\bar{x}_2 - \bar{y}_2)^2}{[-r^{-2}(\bar{x}_1^2 + \bar{x}_2^2) + 1][-r^{-2}(\bar{y}_1^2 + \bar{y}_2^2) + 1]} \right\}^{1/2} \tag{8.84}$$

taken positive.

There is one notable property of 8.84, namely, that $d(\bar{X}\bar{Y}) \to \infty$ as $\bar{Y}$ approaches the absolute $\mathcal{A}$, for any point $\bar{X}$ within $\mathcal{A}$. Thus points outside $\mathcal{A}$ are *inaccessible.* As before, the Euclidean metric is also a limiting case of the hyperbolic metric as $r \to \infty$.

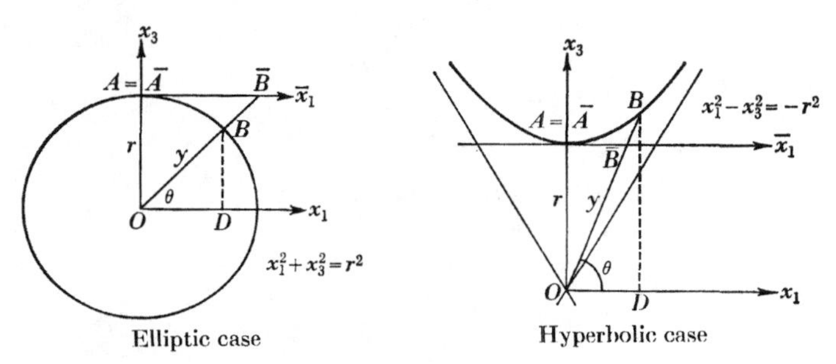

FIG. 8.13

Though the interpretation of the distance function 8.84 as the length of a geodesic on the hyperboloid is no longer valid, by a slight change of emphasis in the elliptic case we can obtain an interpretation which is consistent here. It will be sufficient to consider the case of a plane through the origin intersecting the sphere and the hyperboloid as indicated in Figure 8.13. Instead of defining the length $\overline{AB}$ in terms of the length AB we define it in terms of the *area* of the sector OAB. In the elliptic case the parametric equations of the circle are $x_1 = x = r \cos \theta$, $x_3 = y = r \sin \theta$, so that the area

$$\begin{aligned} OABD &= \int y \, dx = r^2 \int_{\pi/2-\varphi}^{\pi/2} \sin^2\theta \, d\theta = \tfrac{1}{2}r^2 \int_{\pi/2-\varphi}^{\pi/2} (1 - \cos 2\theta) \, d\theta \\ &= \tfrac{1}{2}r^2\varphi + \Delta OBD \end{aligned} \tag{8.85}$$

and in the hyperbolic case $x_1 = x = r \cosh \theta$ and $x_3 = y = r \sinh \theta$, so that the corresponding area

$$\begin{aligned} OABD &= \int y \, dx = -r^2 \int_{\pi/2-\varphi}^{\pi/2} \sinh^2\theta \, d\theta \\ &= \tfrac{1}{2}r^2 \int_{\pi/2-\varphi}^{\pi/2} (1 - \cosh 2\theta) \, d\theta = \tfrac{1}{2}r^2\varphi + \Delta OBD \end{aligned} \tag{8.86}$$

where $d \cosh \theta = \sinh \theta \, d\theta$ and $\cosh 2\theta = 1 + 2 \sinh^2\theta$, as may be easily verified.

Thus in both the elliptic and the hyperbolic case the area of the segment $\widehat{OAB}$ is $\frac{1}{2}r^2\varphi$, and

$$d(\overline{AB}) = \frac{2}{r} \cdot \widehat{OAB}$$

interpreted in the elliptic case by 8.77 and in the hyperbolic case by 8.84. The factor of proportionality $2/r$ is fixed in a given geometry.

In order to obtain the formulas of *hyperbolic trigonometry* we must make the appropriate changes in 8.11, 8.12, and 8.21, which yield

8.871 $$\cosh a = \cosh b \cosh c - \sinh b \sinh c \cos A$$

8.872 $$\frac{\sin A}{\sinh a} = \frac{\sin B}{\sinh b} = \frac{\sin C}{\sinh c}$$

8.873 $$\cos B = -\cos B \cos C + \sin B \sin C \cosh a$$

Thus we are led to a model of the *hyperbolic* plane in which we limit attention to the interior of the absolute circle $\mathcal{A}$. As we have seen, any point on $\mathcal{A}$ is to be considered "*at infinity*." Through X can be drawn *two* parallels XY_∞, XZ_∞ to any line l not passing through X, as well as any number of *ultraparallels* which intersect l in *inaccessible* points P outside $\mathcal{A}$. Thus two lines may intersect, be parallel, or they may not intersect at all in hyperbolic geometry.

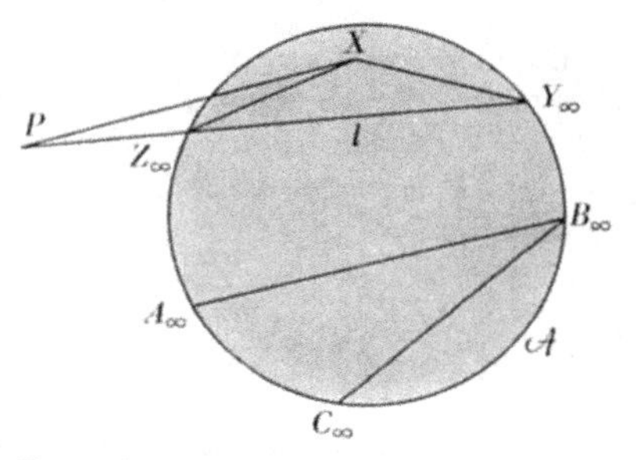

FIG. 8.14

It is beyond the scope of this brief treatment to give a proper definition of "angle" in hyperbolic geometry, but we can say that *perpendicularity* is definable in terms of the harmonic property:

8.88 *Definition* If l is the polar line of an inaccessible point L with regard to the absolute $\mathcal{A}$, then any line "through" L is perpendicular to l. It follows after a little argument, which we omit, that:

(i) Intersecting lines in hyperbolic geometry have no common perpendicular. In elliptic geometry every pair of lines *has* a common perpendicular.

(ii) Parallel lines have no common perpendicular in hyperbolic geometry.

(iii) Nonintersecting lines have one common perpendicular, namely, the polar line of their inaccessible point of intersection.

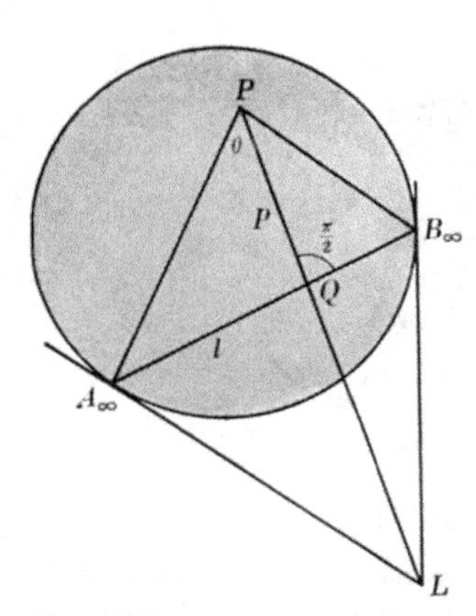

FIG. 8.15

Consider now the triangle PQA_∞ in which PA_∞ is parallel to l and $\angle PQA_\infty = \pi/2$ as in Figure 8.15. If we set $\angle QPA_\infty = \theta$ and $d(PQ) = p$ in 8.73, we conclude that

8.89 $$\sin\theta \cosh p = 1$$

Thus the perpendicular distance p from P to l is determined by and determines the angle θ, called the *angle of parallelism.* It can be shown that the locus of P, such that p remains fixed, is a conic having double contact with $\mathfrak{a}$ at A_∞ and B_∞. Such a curve is called an *equidistant curve.* Since all perpendiculars PQ pass through L, and equidistant curves become parallel lines as $r \to \infty$, we conclude that there is an infinite number of common perpendiculars to two parallel lines in Euclidean geometry, and these are themselves all parallel.

Thus *Euclidean geometry is a limiting case of both elliptic and hyperbolic geometry, and shares its properties in part with one and in part with the other.*

EXERCISES

1. Construct a triangle in the hyperbolic plane whose angle sum is zero.
2. How many such triangles exist having a given line as side?
3. Prove that $\angle QPA_\infty = \angle QPB_\infty$ in Figure 15.
4. Prove that $\theta \to \pi/2$ as $P \to Q$ along QL in Figure 15, by considering (a) the area of the triangle PQA_∞, (b) the relation 8.89.

9 Reduction of Real Matrices to Diagonal Form

9.1 INTRODUCTION

In this our final chapter, we emphasize two important ideas. In the first place, the linear transformation $X' = PX$ is capable of two interpretations. Either it can, when applied to a vector space $\mathcal{V}_n$, be interpreted as a *mapping* of the space on itself in which a vector X is "moved" to a new position X' referred to the same basis vectors; or it can be interpreted as a *renaming* of the fixed vector X relative to a new system of basis vectors $E_i' = PE_i$. These two points of view are complementary, and sometimes it is convenient to adopt one, sometimes the other.

The second important idea is the significance of the *characteristic vectors* of a linear transformation $Y = AX$. In Sections 9.3 and 9.4 we study the problem in the vector space $\mathcal{V}_n$ and in the corresponding projective space $\mathcal{P}_{n-1}$; in Section 9.6 we see when a knowledge of these characteristic vectors of A can determine a matrix P such that PAP^{-1} is diagonal.

If A is *symmetric*, P can be chosen to be orthogonal and this leads in Sections 9.7 and 9.8 to the reduction of a general quadratic polynomial equation to normal form. We explain this reduction process against the background of Klein's Erlanger Programm in an attempt to show the significance of Section 9.5 in which the reduction by elementary transformation given in Chapter 3 is applied to symmetric matrices.

Though these ideas are not easy it is the author's hope that, by confining attention to real matrices and transformations and by continually emphasizing their geometrical importance in an algebraic context, the student may see their underlying meaning. To this end, numerous illustrative examples are worked out in the text. If this meaning can once be grasped, much mathematics studied later and most applications to modern physics and chemistry will become clear.

9.2 CHANGE OF BASIS

In order to combine the two interpretations which can be put upon a linear transformation, let us assume that $|A| \neq 0$ and that the vector

9.21 $$Y = AX$$

is referred to the same basis vectors E_i $(i = 1,2, \ldots n)$ of $\mathcal{V}_n$, while for $|P| \neq 0$,

9.22 $$X' = PX$$

maps the whole space $\mathcal{V}_n$ upon itself, in particular the vectors E_i upon the vectors $PE_i = E_i'$. As we saw in 3.78, E_i' is the ith column vector of the matrix P.

If we multiply 9.21 by the $n \times n$ matrix P, we have

9.231 $$PY = PAX = PAP^{-1}PX$$

and substituting from 9.22 we obtain

9.232 $$Y' = PAP^{-1}X'$$

referred to the new basis vectors E_i' $(i = 1,2, \ldots n)$. Note that the *matrix* of the transformation is now PAP^{-1}.

The requirement that $|P| \neq 0$ is important, since we want the E_i' to be linearly independent vectors in the transformed space which we shall denote by $\mathcal{V}_n'$. But if, in addition, P is *orthogonal*, then

9.241 $$P^tP = P^{-1}P = I$$

which implies that

9.242 $$E_i'^t E_j' = \begin{cases} 1, & i = j \\ 0, & i \neq j \end{cases}$$

and the basis E_i' is *orthonormal* in $\mathcal{V}_n'$. Conversely, the condition 9.241 is a consequence of 9.242, as in 6.17. We illustrate these simple but important ideas in the following examples.

9.25 *Example.* Let us take basis vectors E_1, E_2 as in Figure 9.1, and also

9.251 $$A = \begin{pmatrix} \cos\varphi & -\sin\varphi \\ \sin\varphi & \cos\varphi \end{pmatrix}$$

with

$$P = \begin{pmatrix} \cos\theta & -\sin\theta \\ \sin\theta & \cos\theta \end{pmatrix}$$

so that

9.252 $$Y' = PAP^{-1}X' = AX'$$

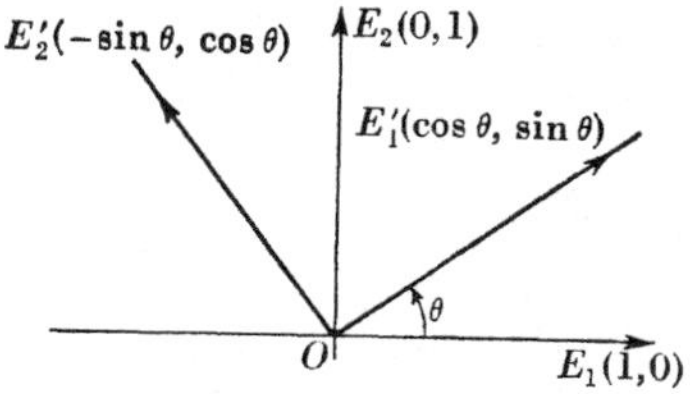

$E_2'(-\sin\theta, \cos\theta)$ $E_2(0,1)$ $E_1'(\cos\theta, \sin\theta)$ θ O $E_1(1,0)$

FIG. 9.1

Since a rotation about the origin is still a rotation through the same angle whether referred to E_1, E_2, or E_1', E_2', P and A *commute* and $PAP^{-1} = A$, as indicated in 9.252.

9.26 *Example.* The situation is quite different in space. Consider first the simple reflection 4.5

9.261 $$Y = \begin{pmatrix} -1 & 0 & 0 \\ 0 & 1 & 0 \\ 0 & 0 & 1 \end{pmatrix} X$$

with

$$P = \begin{pmatrix} 0 & 1 & 0 \\ 1 & 0 & 0 \\ 0 & 0 & 1 \end{pmatrix}$$

so that

9.262 $$Y' = PAP^{-1}X' = \begin{pmatrix} 1 & 0 & 0 \\ 0 & -1 & 0 \\ 0 & 0 & 1 \end{pmatrix} X'$$

and $E_1' = E_2$, $E_2' = E_1$, $E_3' = E_3$. Clearly the geometric transformation itself remains unchanged; it is only described as a reflection in the plane $x_2' = 0$ instead of in the plane $x_1 = 0$.

9.27 *Example.* To take a more complicated case, let us suppose that

9.271 $$Y = \begin{pmatrix} 0 & 1 & 0 \\ 1 & 0 & 0 \\ 0 & 0 & 1 \end{pmatrix} X$$

is transformed by

$$X' = PX = \begin{pmatrix} 1 & 0 & 1 \\ 0 & -1 & 3 \\ 1 & 0 & 2 \end{pmatrix} X$$

$$X = P^{-1}X' = \begin{pmatrix} 2 & 0 & -1 \\ -3 & -1 & 3 \\ -1 & 0 & 1 \end{pmatrix} X'$$

so that (cf. the example of Section 4.5)

$$Y' = \begin{pmatrix} 1 & 0 & 1 \\ 0 & -1 & 3 \\ 1 & 0 & 2 \end{pmatrix} \begin{pmatrix} 0 & 1 & 0 \\ 1 & 0 & 0 \\ 0 & 0 & 1 \end{pmatrix} \begin{pmatrix} 2 & 0 & -1 \\ -3 & -1 & 3 \\ -1 & 0 & 1 \end{pmatrix} X'$$

9.272
$$= \begin{pmatrix} -4 & -1 & 4 \\ -5 & 0 & 4 \\ -5 & -1 & 5 \end{pmatrix} X'$$

Again, the geometrical property of reflection is unchanged, but on substitution from $X = P^{-1}X'$ the equation of the plane $x_1 = x_2$ becomes

$$2x_1' - x_3' = -3x_1' - x_2' + 3x_3'$$

or

9.273
$$5x_1' + x_2' - 4x_3' = 0$$

with $E_1' = (1,0,1)$, $E_2' = (0,-1,0)$, $E_3' = (1,3,2)$. In order to verify the fact that we still have a reflection in the plane 9.273, we note that

9.274
$$\begin{pmatrix} -4 & -1 & 4 \\ -5 & 0 & 4 \\ -5 & -1 & 5 \end{pmatrix} \begin{pmatrix} a_1 \\ a_2 \\ a_3 \end{pmatrix} = \begin{pmatrix} a_1 \\ a_2 \\ a_3 \end{pmatrix}$$

if and only if the vector (a_1,a_2,a_3) lies in the plane 9.273.

Finally, we draw attention to the fact that besides defining $E_i' = PE_i$ as the ith column of the matrix P, we could write:

9.28
$$E_i' = (p_{1i}, p_{2i}, \ldots, p_{ni}) = p_{1i}E_1 + p_{2i}E_2 + \ldots + p_{ni}E_n$$

These relations express the linear dependence of the new basis on the old, and provide another way of describing the matrix P. One should be careful, however, to note that the matrix of coefficients of the E_i in 9.28 is P^t and *not* P.

EXERCISES

1. Transform the matrix A of 9.272 by the matrix

$$P = \begin{pmatrix} 0 & 0 & 1 \\ 1 & 0 & 0 \\ 0 & 1 & 0 \end{pmatrix}$$

What is the equation of the locus of the fixed points of the transformation $Y = PAP^{-1}X$? Verify your conclusion as in 9.274 above.

2. If the new basis elements are defined by the equations

$$\begin{aligned} E_1' &= E_1 - E_2 + E_3 \\ E_2' &= E_1 + E_2 - E_3 \\ E_3' &= -E_1 + E_2 + E_3 \end{aligned}$$

write the equations of the corresponding transformation $X' = PX$ and express X' linearly in terms of the E_i'.

9.3 CHARACTERISTIC VECTORS

First, we observe that in the case of 9.261 we are reflecting in the plane $x_1 = 0$ and this becomes $x_1' = x_2'$, by applying the transformations

$$X' = \begin{pmatrix} \frac{1}{2}\sqrt{2} & \frac{1}{2}\sqrt{2} & 0 \\ -\frac{1}{2}\sqrt{2} & \frac{1}{2}\sqrt{2} & 0 \\ 0 & 0 & 1 \end{pmatrix} X, \qquad X = \begin{pmatrix} \frac{1}{2}\sqrt{2} & -\frac{1}{2}\sqrt{2} & 0 \\ \frac{1}{2}\sqrt{2} & \frac{1}{2}\sqrt{2} & 0 \\ 0 & 0 & 1 \end{pmatrix} X'$$

Moreover,

$$\begin{pmatrix} \frac{1}{2}\sqrt{2} & \frac{1}{2}\sqrt{2} & 0 \\ -\frac{1}{2}\sqrt{2} & \frac{1}{2}\sqrt{2} & 0 \\ 0 & 0 & 1 \end{pmatrix} \begin{pmatrix} -1 & 0 & 0 \\ 0 & 1 & 0 \\ 0 & 0 & 1 \end{pmatrix} \begin{pmatrix} \frac{1}{2}\sqrt{2} & -\frac{1}{2}\sqrt{2} & 0 \\ \frac{1}{2}\sqrt{2} & \frac{1}{2}\sqrt{2} & 0 \\ 0 & 0 & 1 \end{pmatrix} = \begin{pmatrix} 0 & 1 & 0 \\ 1 & 0 & 0 \\ 0 & 0 & 1 \end{pmatrix}$$

so that 9.261 and 9.271 are *geometrically* the same. We have seen that 9.271 and 9.272 are *geometrically* the same, so that all three operations are *geometrically* the same, though they are described with respect to different basis vectors. These basis vectors are orthogonal in the case of 9.261 and 9.271 but not in the case of 9.272.

The important question which we ask now is, *how could we deduce the geometrical identity of the three transformations 9.261, 9.271, 9.272 from their matrices alone?* The answer is provided by considering the nature of their *fixed elements*. Such a line of thought was significant for the inversion transformation in Section 8.5, and we follow it again here. The only difference is that in this case we are dealing with *vectors* and it will be sufficient to ask that the *direction* remain fixed, though the *magnitude* may vary. With this in mind we look for the solutions of the three equations

9.31 $$\lambda X = \begin{pmatrix} -1 & 0 & 0 \\ 0 & 1 & 0 \\ 0 & 0 & 1 \end{pmatrix} X, \quad \text{(a)} \qquad \lambda X = \begin{pmatrix} 0 & 1 & 0 \\ 1 & 0 & 0 \\ 0 & 0 & 1 \end{pmatrix} X, \quad \text{(b)}$$

$$\lambda X = \begin{pmatrix} -4 & -1 & 4 \\ -5 & 0 & 4 \\ -5 & -1 & 5 \end{pmatrix} X \quad \text{(c)}$$

Since the corresponding scalar equations are homogeneous, they have non-trivial solutions if and only if

9.32
$$\begin{vmatrix} \lambda+1 & 0 & 0 \\ 0 & \lambda-1 & 0 \\ 0 & 0 & \lambda-1 \end{vmatrix} = 0, \qquad \begin{vmatrix} \lambda & -1 & 0 \\ -1 & \lambda & 0 \\ 0 & 0 & \lambda+1 \end{vmatrix} = 0,$$

$$\begin{vmatrix} \lambda+4 & 1 & -4 \\ 5 & \lambda & -4 \\ 5 & 1 & \lambda-5 \end{vmatrix} = 0$$

and *each of these three equations reduces to*

9.33
$$(\lambda + 1)(\lambda - 1)(\lambda - 1) = 0$$

In order to treat the problem in general, we note that the matrix equation $\lambda \mathbf{X} = A\mathbf{X}$ is always equivalent to a set of homogeneous scalar equations, and for consistency we must have

9.34
$$|\lambda I - A| = 0$$

This is called the *characteristic equation* of A. But we also have the matrix equation

$$P(\lambda I - A)P^{-1} = \lambda I - PAP^{-1}$$

and taking determinants,

$$|P||\lambda I - A||P^{-1}| = |\lambda I - PAP^{-1}| = |\lambda I - A| = 0$$

From the identity of the characteristic equations of A and PAP^{-1} we conclude that

9.35 *The roots of the characteristic equation of the matrix A are the same as those of the characteristic equation of the matrix PAP^{-1}.*

We call these roots the *characteristic roots* of A, or the *eigenvalues* of A. The matrices A and PAP^{-1} are said to be *similar.*

In order to interpret this algebraic result geometrically we return to the matrix equation $\lambda \mathbf{X} = A\mathbf{X}$ and find a *characteristic* or *eigen vector* $\mathbf{X}_i$ which corresponds to the characteristic root λ_i. Let us consider our previous examples and, in particular, the three equations 9.31.

In 9.31 (a), setting $\lambda = +1$ we conclude that any vector satisfying the condition $x_1 = -x_1$, that is, $x_1 = 0$, will remain fixed under 9.261. All such vectors lie in the *plane* $x_1 = 0$. On the other hand if we set $\lambda = -1$, then the only solution of 9.31 (a) is of the form $(k,0,0)$ which is a vector normal to $x_1 = 0$.

In 9.31 (b), setting $\lambda = +1$ we obtain the condition $x_1 = x_2$ which is the *form* the equation $x_1 = 0$ takes with reference to the basis

$$E_1' = \left(\frac{1}{\sqrt{2}}, -\frac{1}{\sqrt{2}}, 0\right), \quad E_2' = \left(\frac{1}{\sqrt{2}}, \frac{1}{\sqrt{2}}, 0\right), \quad E_3' = (0,0,1)$$

Again, the characteristic root $\lambda = -1$ leads to a solution $(k,-k,0)$ which is the corresponding characteristic vector normal to the plane $x_1 = x_2$.

Finally, if we set $\lambda = +1$ in 9.31 (c) we obtain the equations

9.36
$$\begin{aligned} -4x_1 - x_2 + 4x_3 &= x_1 \\ -5x_1 + 4x_3 &= x_2 \\ -5x_1 - x_2 + 5x_3 &= x_3 \end{aligned}$$

which all coincide with 9.273. Setting $\lambda = -1$,

9.37
$$\begin{aligned} -4x_1 - x_2 + 4x_3 &= -x_1 \\ -5x_1 + 4x_3 &= -x_2 \\ -5x_1 - x_2 + 5x_3 &= -x_3 \end{aligned}$$

and these equations have as solution space the vector (k,k,k), which is the way the normal to the plane $x_1 = x_2$ is described relative to the new basis. This is easily verified by transforming the vector $(1,-1,0)$ according to

9.38
$$\begin{pmatrix} 1 \\ 1 \\ 1 \end{pmatrix} = \begin{pmatrix} 1 & 0 & 1 \\ 0 & -1 & 3 \\ 1 & 0 & 2 \end{pmatrix} \begin{pmatrix} 1 \\ -1 \\ 0 \end{pmatrix}$$

Note that perpendicularity is no longer easily recognizable, since the transformation 9.38 is not orthogonal.

We draw the obvious conclusion from these illustrative examples that *it would be highly desirable if we could confine our attention to orthogonal transformations only.* Though this is not always possible, we shall see shortly how much can be accomplished with such a restriction.

EXERCISES

1. Determine the characteristic roots and the characteristic vectors of the linea transformation

$$Y = \begin{pmatrix} 2 & 0 & -1 \\ 0 & 2 & 0 \\ -1 & 0 & 2 \end{pmatrix} X$$

and explain their geometrical significance.

2. Make the change of variables in Exercise 1 indicated by the transformation

$$X' = PX = \begin{pmatrix} 1 & 0 & 1 \\ 0 & -1 & 3 \\ 1 & 0 & 2 \end{pmatrix} X$$

(cf. the example of Section 3.5) and verify that the characteristic roots remain unchanged. What are the characteristic vectors?

3. Euler's transformation for a rotation through an angle θ about an axis with direction cosines l_1, l_2, l_3 can be written

$$X' = \begin{pmatrix} \cos\theta + & -l_3 \sin\theta + & l_2 \sin\theta + \\ l_1^2(1 - \cos\theta), & l_1l_2(1 - \cos\theta), & l_1l_3(1 - \cos\theta) \\ l_3 \sin\theta + & \cos\theta + & -l_1 \sin\theta + \\ l_2l_1(1 - \cos\theta), & l_2^2(1 - \cos\theta), & l_2l_3(1 - \cos\theta) \\ -l_2 \sin\theta + & l_1 \sin\theta + & \cos\theta + \\ l_3l_1(1 - \cos\theta), & l_3l_2(1 - \cos\theta), & l_3^2(1 - \cos\theta) \end{pmatrix} X$$

Verify that any point on the axis remains fixed.

4. Obtain the various rotations of the cube in Section 4.5 as special cases of Euler's transformation in Exercise 3.

5. Obtain the various rotations of the cube in Section 4.5 as special cases of Rodrigues' transformation in Exercise 8 of Section 6.1

9.4 COLLINEATIONS

As we have done so often before, we can interpret the components of a 3-vector as the homogeneous coordinates of a point in the plane. Since the equation of a line l is homogeneous in $x_1, x_2, x_3 = x_0$, for $n = 3$ the linear transformation $Y = AX$ can be thought of as transforming the line l into a line l' described with reference to the same basis vectors. With this interpretation, we ask what points remain fixed under such a *collineation*? Since $X = \lambda X$ in homogeneous coordinates, as in Section 7.2, this is just the same question we asked before and leads to the characteristic equation

$$|\lambda I - A| = 0 = \begin{vmatrix} \lambda - a_{11} & -a_{12} & -a_{13} \\ -a_{21} & \lambda - a_{22} & -a_{23} \\ -a_{31} & -a_{32} & \lambda - a_{33} \end{vmatrix}$$

which, when expanded, yields the cubic

9.41 $$\lambda^3 + c_1\lambda^2 + c_2\lambda + c_3 = 0$$

with real coefficients. Since 9.41 has three roots $\lambda_1, \lambda_2, \lambda_3$, there are the following possibilities:

9.42 $\lambda_1 = \lambda_2 = \lambda_3$. In this case every point in the plane remains fixed and the collineation is the identity transformation in the plane.

9.43 $\lambda_1 = \lambda_2 \neq \lambda_3$. In the Examples 9.25, 9.26, 9.27 the cubic 9.41 takes the form 9.33 and the transformations are called *homologies*. In the case of 9.261 the *center* of the homology is the point (1,0,0) while the *axis* of the homology is the line $x_1 = 0$. Every point $(0,x_2,x_3)$ of the axis remains fixed under 9.261, as also does every line through the center. To see this last statement it is only necessary to observe that such a line PS has equation

$$\lambda x_2 + \mu x_3 = 0$$

which remains unchanged by 9.261, though a point $(1,\mu,-\lambda)$ will be transformed into the different point $(-1,\mu,-\lambda)$, also on PS.

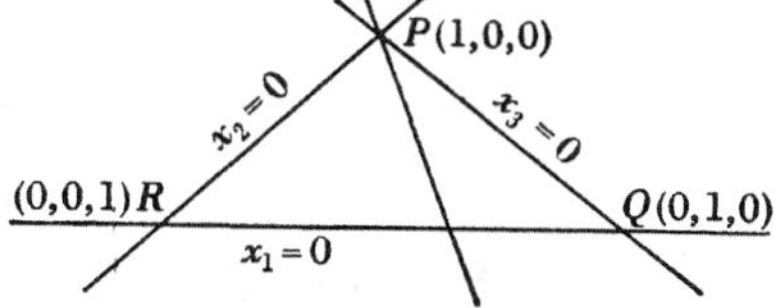

FIG. 9.2

9.44 $\lambda_1 \neq \lambda_2 \neq \lambda_3$. The collineation

$$Y = \begin{pmatrix} a & 0 & 0 \\ 0 & b & 0 \\ 0 & 0 & c \end{pmatrix} X$$

with a, b, c real and different, leaves fixed each of $P(1,0,0)$, $Q(0,1,0)$, and $R(0,0,1)$, and the lines joining these points, but no other points remain fixed. For example, the point $(\lambda,\mu,0)$ on PQ is transformed into the *different* point $(a\lambda,b\mu,0)$ which also lies on PQ.

9.45 $\lambda_1 = \bar{\lambda}_2 \neq \lambda_3$. In order to illustrate the case of only one real root, consider the transformation (cf. Example 9.25)

$$Y = \begin{pmatrix} \cos\theta & -\sin\theta & 0 \\ \sin\theta & \cos\theta & 0 \\ 0 & 0 & 1 \end{pmatrix} X$$

which, in 3-space, is a rotation about the axis $x_1 = x_2 = 0$. The characteristic equation is

$$\begin{vmatrix} \lambda - \cos\theta & \sin\theta & 0 \\ -\sin\theta & \lambda - \cos\theta & 0 \\ 0 & 0 & \lambda - 1 \end{vmatrix} = 0$$

or

$$(\lambda^2 - 2\cos\theta + 1)(\lambda - 1) = (\lambda - e^{i\theta})(\lambda - e^{-i\theta})(\lambda - 1) = 0$$

and (0,0,1) is the only fixed point of the transformation.

EXERCISES

Describe the fixed elements of the collineations

$$\text{(a)}\ Y = \begin{pmatrix} 1 & 0 & 0 \\ 0 & 1 & 0 \\ 0 & 0 & 1 \end{pmatrix} X, \qquad \text{(b)}\ Y = \begin{pmatrix} 1 & 0 & 0 \\ 0 & 1 & 1 \\ 0 & 0 & 1 \end{pmatrix} X$$

$$(c)\ Y = \begin{pmatrix} 1 & 1 & 0 \\ 0 & 1 & 1 \\ 0 & 0 & 1 \end{pmatrix} X, \qquad (d)\ Y = \begin{pmatrix} 1 & 0 & 0 \\ 0 & 1 & 0 \\ 0 & 0 & a \end{pmatrix} X$$

$$(e)\ Y = \begin{pmatrix} 1 & 1 & 0 \\ 0 & 1 & 0 \\ 0 & 0 & a \end{pmatrix} X$$

9.5 REDUCTION OF A SYMMETRIC MATRIX

In Section 3.4 we studied the reduction of an arbitrary matrix A to canonical form PAQ by multiplying on the left and on the right with products of elementary matrices. The particular case in which A is *symmetric*, i.e., in which $a_{ij} = a_{ji}$ $(i,j = 1,2, \ldots, n)$, is important, since then we may perform the *same* operation on rows and columns by multiplying on the left by, say, P_i, and on the right by $Q_i = P_i^t$. In such a case

$$P = P_s \ldots P_2 P_1, \qquad Q = P^t = P_1^t P_2^t \ldots P_s^t \tag{9.51}$$

and *we can choose P so that PAP^t is diagonal.* Moreover,

$$(PAP^t)^t = (P^t)^t A^t P^t = PAP^t \tag{9.52}$$

so that PAP^t remains symmetric at every stage of the process.

Example. Consider the symmetric matrix

$$A = \begin{pmatrix} 1 & 1 & -1 \\ 1 & 2 & 0 \\ -1 & 0 & 1 \end{pmatrix} = A^t$$

Evidently

$$\begin{pmatrix} 1 & 0 & 0 \\ -1 & 1 & 0 \\ 0 & 0 & 1 \end{pmatrix} A \begin{pmatrix} 1 & -1 & 0 \\ 0 & 1 & 0 \\ 0 & 0 & 1 \end{pmatrix} = \begin{pmatrix} 1 & 0 & -1 \\ 0 & 1 & 1 \\ -1 & 1 & 1 \end{pmatrix}$$

and similarly

$$\begin{pmatrix} 1 & 0 & 0 \\ 0 & 1 & 0 \\ 1 & 0 & 1 \end{pmatrix} \begin{pmatrix} 1 & 0 & -1 \\ 0 & 1 & 1 \\ -1 & 1 & 1 \end{pmatrix} \begin{pmatrix} 1 & 0 & 1 \\ 0 & 1 & 0 \\ 0 & 0 & 1 \end{pmatrix} = \begin{pmatrix} 1 & 0 & 0 \\ 0 & 1 & 1 \\ 0 & 1 & 0 \end{pmatrix}$$

so that

$$\begin{pmatrix} 1 & 0 & 0 \\ 0 & 1 & 0 \\ 0 & -1 & 1 \end{pmatrix} \begin{pmatrix} 1 & 0 & 0 \\ 0 & 1 & 1 \\ 0 & 1 & 0 \end{pmatrix} \begin{pmatrix} 1 & 0 & 0 \\ 0 & 1 & -1 \\ 0 & 0 & 1 \end{pmatrix} = \begin{pmatrix} 1 & 0 & 0 \\ 0 & 1 & 0 \\ 0 & 0 & -1 \end{pmatrix}$$

In this case:

$$P = \begin{pmatrix} 1 & 0 & 0 \\ 0 & 1 & 0 \\ 0 & -1 & 1 \end{pmatrix} \begin{pmatrix} 1 & 0 & 0 \\ 0 & 1 & 0 \\ 1 & 0 & 1 \end{pmatrix} \begin{pmatrix} 1 & 0 & 0 \\ -1 & 1 & 0 \\ 0 & 0 & 1 \end{pmatrix} = \begin{pmatrix} 1 & 0 & 0 \\ -1 & 1 & 0 \\ 2 & -1 & 1 \end{pmatrix}$$

The chief reason we are so interested in symmetric matrices A is that every homogeneous quadratic form can be written $\mathbf{X}^t A\mathbf{X}$, so that the equation of any conic, quadric, etc., can be written

9.53 $$\mathbf{X}^t A\mathbf{X} + \text{linear terms} + \text{constant} = 0$$

in nonhomogeneous coordinates, or in the form $\mathbf{X}^t A\mathbf{X} = 0$ in homogeneous coordinates (cf. 7.25 and 7.26). If we make the change of basis represented by the linear transformation $\mathbf{X} = P^t\mathbf{X}'$, the quadratic form $\mathbf{X}^t A\mathbf{X}$ becomes

9.54 $$\mathbf{X}'^t(PAP^t)\mathbf{X}'$$

and we have seen how to choose P so that PAP^t is diagonal. Thus we can calculate the rank of A, which is also called the *rank* of the quadratic form $\mathbf{X}^t A\mathbf{X}$, and so classify conics, quadrics, etc., relative to this invariant.

EXERCISES

1. Write the equations of the conics 6.21, 6.22, 6.24 in homogeneous form $\mathbf{X}^t A\mathbf{X} = 0$, and calculate the *rank* of A in each case.
2. Consider the conics in Exercise 1 as cylinders in space—how does this affect the rank of A?
3. Calculate the rank of A for each of the quadrics 6.31–6.36 when these are written in homogeneous form.
4. Calculate the rank of A if $\mathbf{X}^t A\mathbf{X} = 0$ represents a pair of coplanar lines which (a) intersect, (b) are parallel, (c) coincide.

9.6 SIMILAR MATRICES

As we saw in 9.35, PAP^{-1} and A have the same characteristic equations and the same characteristic roots but *different* characteristic vectors. In fact, if $\mathbf{X}$ is characteristic for A with

9.61 $$\lambda\mathbf{X} = A\mathbf{X}$$

then

9.62 $$\lambda P\mathbf{X} = PAP^{-1}\cdot P\mathbf{X}$$

and $P\mathbf{X}$ is characteristic for PAP^{-1}. If we assume that PAP^{-1} is diagonal then its characteristic vectors must be the basis vectors $\mathbf{E}_i$, each associated with a particular characteristic root λ_i. It follows that in this case the characteristic vectors of A are $\mathbf{X}_i$, where $\mathbf{E}_i = P\mathbf{X}_i$ or

9.63 $$\mathbf{X}_i = P^{-1}\mathbf{E}_i$$

Thus we conclude that

9.64 *If $|P| \neq 0$ and if PAP^{-1} is diagonal, then the characteristic vectors of A are the column vectors of P^{-1}.*

To state a converse theorem, we must know more than the characteristic roots since if some of these are repeated, various possibilities arise such as are referred to in the exercises at the end of Section 9.4. However, if the characteristic roots are all *distinct* we can make the following statement:

9.65 *If the characteristic roots of A are all distinct, then A is similar to a diagonal matrix.*

Proof. It is only necessary to prove that the n characteristic vectors $\mathbf{X}_i$ corresponding to the n characteristic roots λ_i ($i = 1,2, \ldots, n$) are linearly independent, so that, taking them as the column vectors of P^{-1}, we have a nonsingular matrix.

To this end we have from 9.61 that

9.66
$$\lambda_i\mathbf{X}_i = A\mathbf{X}_i \qquad i = 1,2, \ldots, n$$

If $c_1, c_2, \ldots, c_n$ are constants, not all zero, such that $\Sigma\, c_i\mathbf{X}_i = 0$, then also

9.67
$$\Sigma\, c_i\lambda_i\mathbf{X}_i = A\,\Sigma\, c_i\mathbf{X}_i = 0$$

But the λ_i are all different real numbers so that 9.67 implies that

9.68
$$\Sigma'\, c_i(\lambda_i - \lambda_j)\mathbf{X}_i = 0$$

where Σ' omits the term with $i = j$, and *j can take any value from* 1 *to* n. But this yields a contradiction, since such a relation can be assumed to involve a minimum number of c's which are all different from zero and 9.68 then implies that the number could be reduced further by one. Thus no such relation exists and the $\mathbf{X}_i$ are linearly independent. It follows that the matrix P^{-1} is nonsingular and PAP^{-1} has characteristic vectors $\mathbf{E}_i = P\mathbf{X}_i$, from which we conclude that PAP^{-1} is diagonal with characteristic roots $\lambda_1,\lambda_2, \ldots \lambda_n$.

Example. In order to illustrate these ideas, consider the matrix

$$A = \begin{pmatrix} -1, & 0, & 2 \\ -3, & 2, & 3 \\ -4, & 0, & 5 \end{pmatrix}$$

whose characteristic roots are $\lambda_1 = 1$, $\lambda_2 = 2$, $\lambda_3 = 3$. The corresponding characteristic vectors are obtained by solving the sets of equations

$$\begin{aligned} -x_1 + 2x_3 &= x_1 \\ -3x_1 + 2x_2 + 3x_3 &= x_2 \\ -4x_1 + 5x_3 &= x_3 \end{aligned} \qquad \begin{aligned} -x_1 + 2x_3 &= 2x_1 \\ -3x_1 + 2x_2 + 3x_3 &= 2x_2 \\ -4x_1 + 5x_3 &= 2x_3 \end{aligned}$$

$$\begin{aligned} -x_1 + 2x_3 &= 3x_1 \\ -3x_1 + 2x_2 + 3x_3 &= 3x_2 \\ -4x_1 + 5x_3 &= 3x_3 \end{aligned}$$

Thus $X_1 = (\alpha,0,\alpha)$, $X_2 = (0,\beta,0)$, $X_3 = (\gamma,3\gamma,2\gamma)$, where $\alpha \neq 0$, $\beta \neq 0$, $\gamma \neq 0$, so that

$$P^{-1} = \begin{pmatrix} \alpha & 0 & \gamma \\ 0 & \beta & 3\gamma \\ \alpha & 0 & 2\gamma \end{pmatrix}, \qquad P = \begin{pmatrix} \frac{2}{\alpha} & 0 & -\frac{1}{\alpha} \\ \frac{3}{\beta} & \frac{1}{\beta} & -\frac{3}{\beta} \\ -\frac{1}{\gamma} & 0 & \frac{1}{\gamma} \end{pmatrix}$$

It may be verified that

$$PAP^{-1} = \begin{pmatrix} 1 & 0 & 0 \\ 0 & 2 & 0 \\ 0 & 0 & 3 \end{pmatrix}$$

and the determination of the characteristic vectors has enabled us to make a change of basis, such that PAP^{-1} is diagonal. Note that P is determined up to arbitrary nonzero constants α, β, γ; in other words, the *magnitudes* of the characteristic vectors of A, which are also the basis vectors of PAP^{-1}, are not important.

EXERCISES

1. Using the characteristic vectors of the transformation $Y = AX$ in Exercise 1 of Section 9.3, construct a matrix P such that PAP^{-1} is diagonal. Is the matrix P unique?
2. Construct the most general matrix which commutes with the matrix of collineation (b) in the exercises of Section 9.4 (set $AM = MA$ and equate coefficients).
3. Interpret the most general matrix which commutes with the matrix of collineation (d) in the exercises of Section 9.4 with $a \neq 1$, with reference to the three examples of Section 9.2 (cf. Section 9.3).

9.7 ORTHOGONAL REDUCTION OF A SYMMETRIC MATRIX

In Section 9.5 we saw that the *symmetry* of a matrix about its principal diagonal has certain consequences; a further consequence is the following:

9.71 *The characteristic roots of a real symmetric matrix are all real.*

Proof. Let us assume that $A = A^t$, and if the equation 9.61 has a characteristic root λ and characteristic vector X, then we may assume for the moment that the components $(x_1,x_2, \ldots x_n)$ of X are complex, and denote the conjugate vector by $\overline{X}$ with components $(\overline{X}_1,\overline{X}_2, \ldots \overline{X}_n)$. Thus

$$\lambda X^t\overline{X} = X^tA\overline{X}, \qquad \lambda \overline{X}^tX = \overline{X}^tAX$$

but since

$$X^t\overline{X} = \overline{X}^tX = \overline{X^t\overline{X}}$$

we conclude that

$$X^tA\overline{X} = \overline{X}^tAX = \overline{X^tA\overline{X}}$$

and both $X^t\overline{X}$ and $X^tA\overline{X}$ are real. It follows that

$$\lambda = \overline{X}^tAX/\overline{X}^tX$$

since $\overline{X}^tX \neq 0$, and λ must be real.

The question now arises, does this *reality* of the characteristic roots of a real symmetric matrix A have significance for the process of bringing A to diagonal form? We prove the following important result:

9.72 *If A is a real symmetric matrix, there exists an orthogonal matrix P such that PAP^{-1} is diagonal.*

Proof. Our problem is to utilize the fact that A is symmetric in the argument of the preceding section. Since it has real characteristic roots λ_i, we conclude that the equations

$$\lambda_i X = AX$$

must define *real* characteristic vectors X_i. These vectors need not all be distinct, but we can assume that

$$|X_i| = \{X_iX_i\}^{1/2} = 1$$

as in Section 5.2.

Consider now one of these real characteristic vectors, say X_1, whose components we may suppose to be

$$p_{11}, p_{12}, \ldots p_{1n}$$

Since $\lambda_1X_1 = AX_1$, we conclude that the subspace $\mathcal{V}_{n-1}$ of $\mathcal{V}_n$ orthogonal to X_1 is mapped on itself by A. Thus if we choose an orthonormal basis

$$q_{i1}, q_{i2}, \ldots q_{in} \qquad\qquad i = 2, \ldots n$$

in $\mathcal{V}_{n-1}$, the matrix P_1^{-1} with these as column vectors is *orthogonal:*

$$P_1 = \begin{pmatrix} p_{11} & p_{12} & \cdots & p_{1n} \\ q_{21} & q_{22} & \cdots & q_{2n} \\ \cdot & & & \\ \cdot & & & \\ \cdot & & & \\ q_{n1} & q_{n2} & \cdots & q_{nn} \end{pmatrix}, \qquad P_1^{-1} = \begin{pmatrix} p_{11} & q_{21} & \cdots & q_{n1} \\ p_{12} & q_{22} & \cdots & q_{n2} \\ \cdot & & & \\ \cdot & & & \\ \cdot & & & \\ p_{1n} & q_{2n} & \cdots & q_{nn} \end{pmatrix} = P_1^t$$

It follows from 9.63 that $(1,0,0, \ldots 0)$ is a characteristic vector of PAP^{-1}, and since $PAP^{-1} \cdot PX_1 = \lambda_1 PX_1$,

$$P_1AP_1^{-1} = \begin{pmatrix} \lambda_1 & 0 \\ B_1 & A_1 \end{pmatrix}$$

Since

$$(P_1AP_1^{-1})^t = (P_1AP_1^t)^t = P_1A^tP_1^t = P_1AP_1^{-1}$$

$P_1AP_1^{-1}$ is symmetric, so that $B_1 = 0$ and

$$P_1AP_1^{-1} = \begin{pmatrix} \lambda_1 & 0 \\ 0 & A_1 \end{pmatrix}$$

where A_1 is symmetric.

We have thus established the basis for an induction, since we can similarly construct an orthogonal matrix P_2 such that

$$P_2P_1AP_1^{-1}P_2^{-1} = \begin{pmatrix} 1 & 0 \\ 0 & P_2' \end{pmatrix}\begin{pmatrix} \lambda_1 & 0 \\ 0 & A_1 \end{pmatrix}\begin{pmatrix} 1 & 0 \\ 0 & P_2'^{-1} \end{pmatrix}$$
$$= \begin{pmatrix} \lambda_1 & 0 & 0 \\ 0 & \lambda_2 & 0 \\ 0 & 0 & A_2 \end{pmatrix}$$

where A_2 is symmetric and P_2P_1 is again orthogonal. Proceeding thus, we reach the desired conclusion after at most n steps. Moreover, the matrix of transformation

$$P = P_nP_{n-1} \ldots P_2P_1 = \begin{pmatrix} p_{11} & p_{12} & \cdots & p_{1n} \\ p_{21} & p_{22} & & p_{2n} \\ \vdots & & & \\ p_{n1} & p_{n2} & & p_{nn} \end{pmatrix}$$

is orthogonal, and the columns of $P^{-1} = P^t$ (i.e., the rows of P) are *normal orthogonal characteristic vectors of* A.

EXERCISES

1. Find the characteristic vectors of the symmetric matrix

$$A = \begin{pmatrix} \frac{7}{3} & -\frac{2}{3} & 0 \\ -\frac{2}{3} & 2 & -\frac{2}{3} \\ 0 & -\frac{2}{3} & \frac{5}{3} \end{pmatrix}$$

and construct the orthogonal matrix P such that PAP^{-1} is diagonal (cf. Exercise 9 of Section 6.1).

2. Given that the characteristic roots of the matrix

$$A = \begin{pmatrix} \frac{13}{9} & -\frac{4}{9} & \frac{2}{9} \\ -\frac{4}{9} & \frac{13}{9} & -\frac{2}{9} \\ \frac{2}{9} & -\frac{2}{9} & 1 \end{pmatrix}$$

are 1, 1, 2, carry out the construction of Section 9.7 to obtain an orthogonal matrix P such that PAP^{-1} is diagonal.

9.8 THE REAL CLASSICAL GROUPS

The significance of the notion of a group of linear transformations in geometry goes back to Klein whose Erlanger Programm of 1872 laid the foundations for many of the ideas we have been studying in this book. The following results can easily be verified and are of great importance.

9.81 The totality of all nonsingular real linear transformations on n variables forms a group called the *full linear group* $\mathcal{GL}(n)$. Every such transformation represents a mapping of the vector space $\mathcal{V}_n$ on itself.

9.82 If we interpret the n variables $(x_1, x_2, \ldots x_n)$ as homogeneous coordinates of a point P in a projective space $\mathcal{P}_{n-1}$, then

$$(kx_1, kx_2, \ldots kx_n)$$

represents the same point P for all $k \neq 0$. This freedom in the choice of k enables us to restrict our attention to *unimodular* transformations, i.e., all transformations $\mathbf{Y} = A\mathbf{X}$ having $|A| = 1$ (cf. Section 7.2). Clearly, all such collineations form a subgroup of $\mathcal{GL}(n)$ called the *unimodular group*.

9.83 If we interpret $x_n = x_0 = 0$ as the equation of the *space at infinity* (*line at infinity* for $n = 3$), we have *affine geometry*. If we insist that our unimodular transformation $\mathbf{Y} = A\mathbf{X}$ leave $x_0 = 0$ fixed, it follows that A must have the form

$$\begin{pmatrix} A_0 & T \\ 0 & 1 \end{pmatrix}$$

where A_0 is a unimodular $(n-1) \times (n-1)$ matrix, and T is a column vector having $(n-1)$ components. Since

$$\begin{pmatrix} A_1 & T_1 \\ 0 & 1 \end{pmatrix}\begin{pmatrix} A_2 & T_2 \\ 0 & 1 \end{pmatrix} = \begin{pmatrix} A_3 & T_3 \\ 0 & 1 \end{pmatrix}$$

all such affine transformations form a group called the *affine group* $\mathcal{A}(n-1)$. We can distinguish two important subgroups:

9.84 those transformations in which $T = 0$, and

9.85 those transformations in which $A_0 = I$.

9.86 In the particular case where all the linear transformations $\mathbf{Y} = A\mathbf{X}$ are orthogonal, if $AA^t = 1$, $BB^t = 1$, then $(AB)(AB)^t = ABB^tA^t = 1$, so that we have a subgroup $\mathcal{O}(n)$ of $\mathcal{GL}(n)$, called the *orthogonal group.*

The condition that A and so $|A|$ be real implies that the coefficients of the characteristic equation $|\lambda I - A| = 0$ are real, and further, since $A^tA = I$, that $|A| = \pm 1$. From this we conclude that if λ is a complex characteristic root, then its conjugate $\bar{\lambda}$ is also a characteristic root, and *the corresponding characteristic vectors are complex.* We have an example of this in 9.45, where it can be verified that the characteristic vectors corresponding to $\lambda_1 = e^{i\theta}$, $\lambda_2 = e^{-i\theta}$ are $\mathbf{X}_1 = (i,1)$ and $\mathbf{X}_2 = (-i,1)$. In order to speak of the *magnitude* of a complex vector we must modify slightly our definition, writing $|\mathbf{X}| = \overline{\mathbf{X}}^t\mathbf{X}$, which reduces to $\mathbf{X}^t\mathbf{X}$ if $\mathbf{X}$ is real; but we still have

$$\overline{\mathbf{Y}}^t\mathbf{Y} = \overline{\mathbf{X}}^tA^tA\mathbf{X} = \overline{\mathbf{X}}^t\mathbf{X}$$

if A is orthogonal. It follows that if $\mathbf{Y} = \lambda\mathbf{X}$, where $\mathbf{X}$ is any characteristic vector of the transformation $\mathbf{Y} = A\mathbf{X}$, we must have $\lambda\bar{\lambda} = 1$. Thus

9.861 *The characteristic roots of an orthogonal matrix are roots of unity and, since* $|A| = \pm 1$, *they occur in conjugate complex pairs.*

Just as in the case of the transformation of 9.45, these two conjugate complex characteristic vectors span a real plane, so by suitably choosing the basis vectors we have proved that

9.862 *Any orthogonal* $n \times n$ *matrix* A *can be transformed by an orthogonal matrix* P *so that*

$$PAP^{-1} = \begin{pmatrix} \cos\theta_1 & -\sin\theta_1 & & & & & 0 \\ \sin\theta_1 & \cos\theta_1 & & & & & \\ & & \cos\theta_2 & -\sin\theta_2 & & & \\ & & \sin\theta_2 & \cos\theta_2 & \cdot & & \\ 0 & & & & & \cdot \quad \cdot & \pm 1 \end{pmatrix}$$

We may think of the orthogonal transformation $\mathbf{Y} = A\mathbf{X}$ as consisting of a succession of rotations and (or) reflections performed in mutually orthogonal subspaces. By combining two reflections to make a rotation through π we conclude that

9.863 *Any orthogonal transformation may be considered as a succession of rotations in mutually orthogonal planes if* $|A| = 1$, *followed by a single reflection if* $|A| = -1$.

9.87 If we assume that A_0 in 9.83 is orthogonal we have the *Euclidean group.* For $n = 3$ the situation has been studied in Section 7.5, and the "circular points" 7.52 remain fixed under such a transformation (cf. Exercises 3, 4 of Section 7.5). In nonhomogeneous coordinates, any Euclidean transformation of the plane can be written

9.871
$$\begin{aligned} y_1 &= a_{11}x_1 + a_{12}x_2 + t_1 \\ y_2 &= a_{21}x_1 + a_{22}x_2 + t_2 \end{aligned}$$

where the matrix

$$A_0 = \begin{pmatrix} a_{11} & a_{12} \\ a_{21} & a_{22} \end{pmatrix}$$

is orthogonal (i.e., a rotation or a reflection), and 9.871 can be written in vector form

9.872
$$\mathbf{Y} = A_0\mathbf{X} + T$$

Thus the two subgroups 9.84 and 9.85 of the affine group 9.83 turn out to be, in Euclidean geometry, the group of all orthogonal transformations, and the group of translations in the plane.

Klein's contention was that, while a geometry determines its group of collineations, conversely the group of collineations describes the geometry completely. A "theorem" (such as Pythagoras' theorem) is a relation which remains *invariant* under all collineations of the appropriate group. Since *angle*, *distance*, and so *area*, *volume*, and *generalized volume* are all defined in Euclidean geometry in terms of the inner product $\mathbf{X} \cdot \mathbf{X} = \mathbf{X}^t\mathbf{X}$, it follows from 6.17 and the invariance of $\mathbf{X}^t\mathbf{X}$ under translation that all these are invariant under any Euclidean transformation. This completes the proof of 5.57.

9.9 REDUCTION OF THE GENERAL CONIC TO NORMAL FORM

We conclude our study of vector geometry with a brief analysis of the Euclidean transformations necessary to bring to normal form the general conic (7.241):

9.91
$$ax_1^2 + bx_2^2 + 2hx_1x_2 + 2gx_1 + 2fx_2 + c = 0$$

Two methods are available.

First Method. If we write the quadratic terms of 9.91 in the form

9.92
$$\mathbf{X}^tA_0\mathbf{X} = (x_1, x_2)\begin{pmatrix} a & h \\ h & b \end{pmatrix}\begin{pmatrix} x_1 \\ x_2 \end{pmatrix}$$

and find an orthogonal matrix P such that

$$PA_0P^{-1} = \begin{pmatrix} \lambda_1 & 0 \\ 0 & \lambda_2 \end{pmatrix}$$

as in the preceding section, then setting $X = P^{-1}X'$, 9.91 becomes

9.93
$$\lambda_1 x_1'^2 + \lambda_2 x_2'^2 + 2g'x_1' + 2f'x_2' + c' = 0$$

Completing the squares we have

9.94
$$\lambda_1\left(x_1' + \frac{g'}{\lambda_1}\right)^2 + \lambda_2\left(x_2' + \frac{f'}{\lambda_2}\right)^2 = \frac{g'^2}{\lambda_1} + \frac{f'^2}{\lambda_2} - c'$$

as desired. The method is applicable for any n but we illustrate it in the plane only.

9.95 *Example.* If we take 9.91 to be

9.951
$$x_1^2 + 2x_1x_2 - x_2^2 - 2x_1 + 2x_2 + 3 = 0$$

the quadratic terms can be written X^tA_0X where

9.952
$$A_0 = \begin{pmatrix} 1 & 1 \\ 1 & -1 \end{pmatrix}$$

with the characteristic roots $\lambda_1 = \sqrt{2}$, $\lambda_2 = -\sqrt{2}$ and characteristic vectors $X_1 = (1 + \sqrt{2}, 1)$, $X_2 = (-1, 1 + \sqrt{2})$. Following the construction of the preceding section,

9.953
$$P = \frac{1}{k}\begin{pmatrix} 1 + \sqrt{2} & 1 \\ -1 & 1 + \sqrt{2} \end{pmatrix}, \qquad P^{-1} = \frac{1}{k}\begin{pmatrix} 1 + \sqrt{2} & -1 \\ 1 & 1 + \sqrt{2} \end{pmatrix}$$

where $k = \sqrt{4 + 2\sqrt{2}}$. With the substitution $X = P^{-1}X'$, the equation 9.951 takes the form

$$\sqrt{2}x_1'^2 - \sqrt{2}x_2'^2 - \frac{2\sqrt{2}}{k}x_1' + kx_2' + 3 = 0$$

and completing squares we have

$$\sqrt{2}\left[x_1' - \frac{1}{k}\right]^2 - \sqrt{2}\left[x_2' - \frac{1}{k}(1 + \sqrt{2})\right]^2 = -4$$

If we set

9.954
$$x_1' = x_1'' + \frac{1}{k}$$

$$x_2' = x_2'' + \frac{1}{k}(1 + \sqrt{2})$$

we finally arrive at the normal form

9.955
$$-\frac{x_1''^2}{2\sqrt{2}} + \frac{x_2''^2}{2\sqrt{2}} = 1$$

of a rectangular hyperbola. We sum up the changes of variable in the vector equation

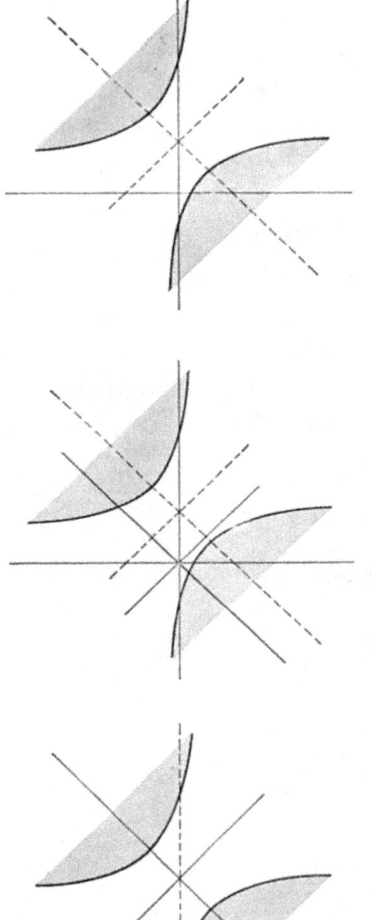

Fig. 9.3

$$9.956 \qquad \mathbf{X} = P^{-1}\mathbf{X}' = P^{-1}(\mathbf{X}'' + \mathbf{C}')$$

or in the scalar equations

$$9.957 \quad x_1 = \frac{1}{k}(1+\sqrt{2})x_1'' - \frac{1}{k}x_2''$$

$$x_2 = \frac{1}{k}x_1'' + \frac{1}{k}(1+\sqrt{2})x_2'' + 1$$

Second Method. If we examine the equation 9.956 we see that $P^{-1}\mathbf{X}''$ represents a *rotation* to bring the coordinate axes parallel to the axes of the conic, while $P^{-1}\mathbf{C}'$ represents a *translation* of the origin to the center $C(0,1)$ of the conic, as is illustrated in Figure 9.3. It is natural to inquire whether we might not profitably reverse this order of procedure and first translate the origin to the center of the conic. If this were done in the case of the above example we would set $\mathbf{X} = \mathbf{Y} + \mathbf{C}$ so that 9.951 would become

$$9.958 \qquad y_1^2 - y_2^2 + 2y_1y_2 + 4 = 0$$

and the subsequent rotation of the axes would involve the quadratic terms only.

The problem resolves itself, then, into finding the center C of the conic in question. This can be accomplished most conveniently with the help of the calculus as we shall explain in the Appendix, but an equivalent procedure is to insist that *the curve be symmetrical with regard to* C, i.e., that its equation remain unchanged when $\mathbf{Y}$ is replaced by $-\mathbf{Y}$. Again referring to the preceding example, this would mean for 9.951 that

$$(y_1 + c_1)^2 + 2(y_1 + c_1)(y_2 + c_2) - (y_2 + c_2)^2 - 2(y_1 + c_1) + 2(y_2 + c_2) + 3$$
$$\equiv (-y_1 + c_1)^2 + 2(-y_1 + c_1)(-y_2 + c_2) - (-y_2 + c_2)^2 - 2(-y_1 + c_1) + 2(-y_2 + c_2) + 3$$

identically, so that

$$4y_1(c_1 + c_2 - 1) + 4y_2(c_1 - c_2 + 1) \equiv 0$$

from which we conclude that

9.959 $$c_1 + c_2 - 1 = 0 = c_1 - c_2 + 1$$

and $c_1 = 0$, $c_2 = 1$ as we expected.

In order to clarify these two approaches to the reduction problem, we give the two corresponding factorizations of the Euclidean transformation, written in homogeneous form, first the direct transformation in 9.96 and then its inverse in 9.97:

9.96 $$\begin{pmatrix} P & -C' \\ 0 & 1 \end{pmatrix} = \begin{pmatrix} I & -C' \\ 0 & 1 \end{pmatrix}\begin{pmatrix} P & 0 \\ 0 & 1 \end{pmatrix} = \begin{pmatrix} P & 0 \\ 0 & 1 \end{pmatrix}\begin{pmatrix} I & -C \\ 0 & 1 \end{pmatrix}$$

(*first method*) (*second method*)

9.97 $$\begin{pmatrix} P^{-1} & C \\ 0 & 1 \end{pmatrix} = \begin{pmatrix} P^{-1} & 0 \\ 0 & 1 \end{pmatrix}\begin{pmatrix} 1 & C' \\ 0 & 1 \end{pmatrix} = \begin{pmatrix} I & C \\ 0 & 1 \end{pmatrix}\begin{pmatrix} P^{-1} & 0 \\ 0 & 1 \end{pmatrix}$$

where $C' = PC$ and C is the center found above. As we have remarked already, the *second method* of procedure is usually preferable.

There remains the possibility that one of the characteristic roots of the matrix A_0 in 9.92 should be zero. But this could happen only if $|A_0| = 0$, in which case we could take $h = ka$, $b = kh = k^2a$ so that not only is the rank of A_0 equal to 1, but when a is divided out *the quadratic terms of 9.91 form a perfect square.* This is the necessary and sufficient condition that 9.91 represent a parabola, and the appropriate Euclidean transformation will bring it to normal form.

Let us summarize our conclusions with regard to the reduction of the quadratic equation

9.98 $$\mathbf{X}^t A \mathbf{X} = \mathbf{X}^t \begin{pmatrix} a & h & g \\ h & b & f \\ g & f & c \end{pmatrix} \mathbf{X} = 0$$

in homogeneous coordinates:

9.981 *If* $|A| = 0$ *the equation 9.98 factors and represents a pair of lines (6.53).*

9.982 *If* $|A| \neq 0$ *and* $|A_0| = \begin{vmatrix} a & h \\ h & b \end{vmatrix} = 0$, *then 9.98 represents a parabola.*

9.983 *If* $|A| \neq 0$ *and* $|A_0| \neq 0$, *then 9.98 represents a central conic which may be an ellipse or an hyperbola.*

This brings us back to Section 9.5 and explains why it is so important to determine the *rank* of A. While this rank of A can be obtained most easily by using elementary transformations, we see now that such transformations are not orthogonal. Consequently they do not preserve the *form* of the curve. We must be clear as to our aims; if it is a case of seeing what a conic looks like, i.e., of plotting it, then we should use only Euclidean transformations so that this form is unchanged.

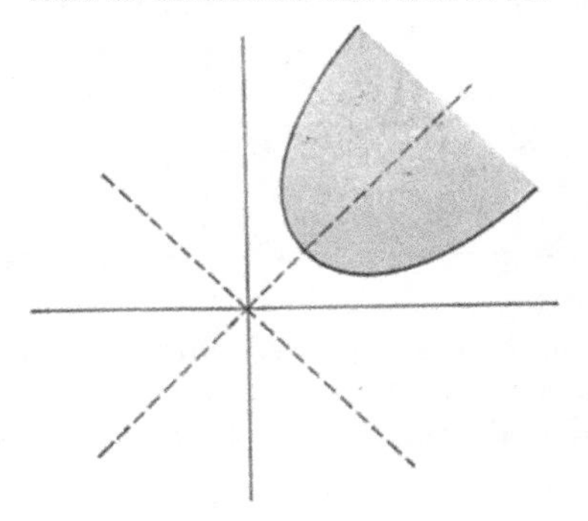

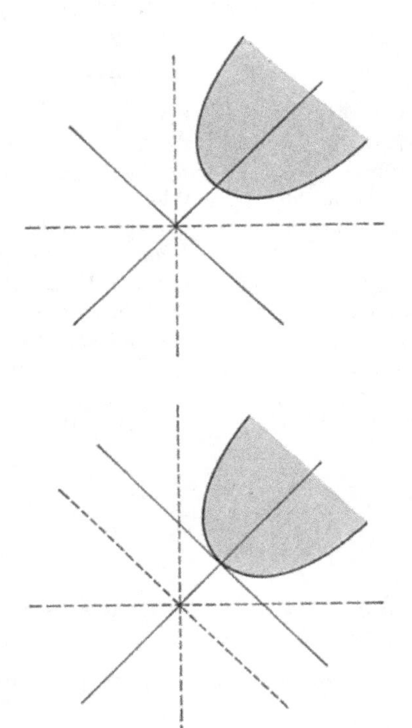

FIG. 9.4

EXERCISES

1. Plot the curve represented by the equation

$$x_1^2 + 2x_1x_2 + x_2^2 - 2x_1 + 2x_2 + 3 = 0$$

Solution. It is easy to verify that $\lambda_1 = 0$ and $\lambda_2 = 2$ so that the corresponding normal characteristic vectors are

$$\left(\frac{1}{\sqrt{2}}, -\frac{1}{\sqrt{2}}\right) \quad \text{and} \quad \left(\frac{1}{\sqrt{2}}, \frac{1}{\sqrt{2}}\right)$$

It follows that

$$P^{-1} = \begin{pmatrix} \frac{1}{\sqrt{2}} & \frac{1}{\sqrt{2}} \\ -\frac{1}{\sqrt{2}} & \frac{1}{\sqrt{2}} \end{pmatrix}$$

and

$$P = \begin{pmatrix} \frac{1}{\sqrt{2}} & -\frac{1}{\sqrt{2}} \\ \frac{1}{\sqrt{2}} & \frac{1}{\sqrt{2}} \end{pmatrix}$$

and substituting $\mathbf{X} = P^{-1}\mathbf{X}'$ we have

$$x_2'^2 = \sqrt{2}\left(x_1' - \frac{3}{2\sqrt{2}}\right)$$

Though the parabola has no center, we translate the origin to the *vertex*

$$V' = PV = \left(\frac{3}{2\sqrt{2}}, 0\right)$$

so that the equation takes the normal form

$$x_2''^2 = \sqrt{2}\, x_1''$$

2. Bring each of the following equations to normal form (i) by the first method, (ii) by the second method, and make diagrams to record the changes of coordinates and the shapes of the corresponding loci.

   (a) $x_1^2 - 3x_1x_2 + 2x_2^2 + x_2 - 1 = 0$
   (b) $2x_1^2 + 4x_1x_2 + 2x_2^2 - x_1 + x_2 + 1 = 0$
   (c) $3x_1^2 - 2x_1x_2 + 3x_2^2 - 4x_1 + 4x_2 + 1 = 0$
   (d) $x_1^2 + 6x_1x_2 + x_2^2 + 2x_1 - 2x_2 = 0$
   (e) $x_1^2 + x_2^2 - x_1 + x_2 + 1 = 0$

3. Bring the quadric

$$x_1^2 + x_1x_2 + x_1x_3 + x_2x_3 + x_1 + x_3 = 0$$

   to the normal form

$$x_1^2 - x_2^2 + x_3^2 = 1$$

   (a) by the first method, (b) by the second method.

4. It can be proved that the quadric represented by the equation 7.26 in homogeneous coordinates:

$$\mathbf{X}^t A \mathbf{X} = 0$$

   is (a) a plane counted twice, if A has rank 1; (b) two distinct planes, if A has rank 2; (c) a cone, if A has rank 3. Construct examples to illustrate each of these three cases.

5. Prove that the center of the quadric represented by the equation

$$f(x_1,x_2,x_3,x_0) \equiv \mathbf{X}^t A \mathbf{X} = 0$$

   is given by the solution of the equations

$$\frac{\partial f}{\partial x_1} = \frac{\partial f}{\partial x_2} = \frac{\partial f}{\partial x_3} = 0$$

   Show that this center is finite if $|a_{ij}| \neq 0$ for $i,j = 1,2,3$.

6. What are the conditions that 7.26 should represent an ellipsoid or hyperboloid?

7. What is the significance of the condition $|a_{ij}| = 0$ for $i,j = 1,2,3$? Give an example.

Appendix: Calculus and Geometry

Like nearly all our mathematical ideas, the calculus had its origin in geometry. In particular, the study of tangents to curves by Fermat and the study of motion by Newton led to the development of the calculus in the 17th Century, though the notation we use today is largely due to Leibniz. Conversely, the expansion of a function in a Taylor's series has geometrical applications which are so significant that it seems unfair not to draw attention to them.

Perhaps the simplest of these applications is to a plane curve $\mathcal{C}_n$ given in the form

$$x_2 = f(x_1) \tag{1}$$

where $f(x_1)$ is a polynomial in x_1 of degree n. If we expand the function $f(x_1)$ about $x_1 = a$, we have

$$x_2 = f(a) + \frac{f'(a)}{1!}(x_1 - a) + \frac{f''(a)}{2!}(x_2 - a)^2 + \ldots + \frac{f^{(n)}(a)(x_1 - a)^n}{n!} \tag{2}$$

since the $(n+1)^{th}$ and all higher derivatives vanish. The advantage of writing (1) in the form (2) is that it enables us to *approximate* to $\mathcal{C}_n$ in the neighborhood of the point $A(a,f(a))$. For example, we could ignore the terms in (2) of order higher than the first to obtain

$$x_2 - f(a) = f'(a)(x_1 - a) \tag{3}$$

which is the equation of the *tangent* to the curve at A. If we substitute in (2) we find that the tangent (3) meets the curve in *two* coincident points if $f''(a) \neq 0$ and in *three* coincident points when $f''(a) = 0$ if $f'''(a) \neq 0$; in the latter case A is called a *point of inflection* of $\mathcal{C}_n$.

Example 1. Consider the cubic curve $x_2 = f(x_1) = x_1^3$ so that

$$x_2 = a^3 + \frac{3a^2}{1!}(x_1 - a) + \frac{6a}{2!}(x_1 - a)^2 + \frac{6}{3!}(x_1 - a)^3$$

At the point (a,a^3) the equation of the tangent is

$$x_2 - a^3 = 3a^2(x_1 - a) \tag{4}$$

which meets the curve in *three* coincident points if $a = 0$. The origin is thus a point of inflection with $x_2 = 0$ as the inflectional tangent. The graph of the curve is shown in the figure.

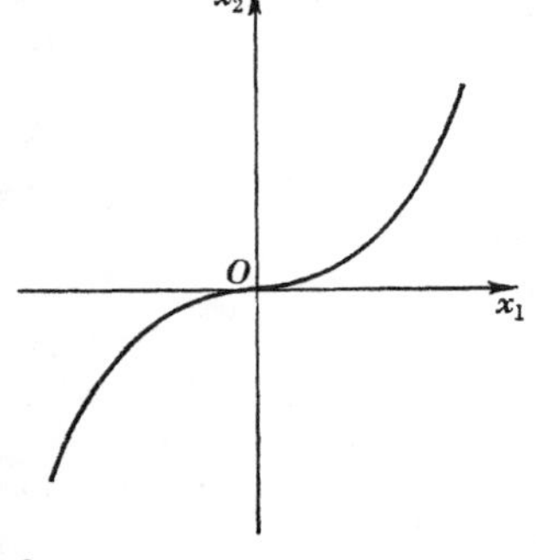

GRAPH

We immediately remark, however, that the condition that $f(x)$ be a polynomial is undesirably restrictive. The best way out of this difficulty is to use homogeneous coordinates and write the equation of the curve $\mathcal{C}_n$ in the form

$$f(x_1,x_2,x_0) = 0$$

where $f(x_1,x_2,x_0)$ is a homogeneous polynomial of degree n in x_1, x_2, x_0. Fortunately, Taylor's theorem generalizes to any number of variables, though we must use partial differentiation to obtain the desired expansion. Let us suppose that the point $X(x_1,x_2,x_0)$ lies on the curve $\mathcal{C}_n$ and that $Y(y_1,y_2,y_0)$ does *not* lie on $\mathcal{C}_n$. Then any intersection of XY with $\mathcal{C}_n$ has homogeneous coordinates

$$(\lambda x_1 + \mu y_1, \quad \lambda x_2 + \mu y_2, \quad \lambda x_0 + \mu y_0) \tag{5}$$

where $\lambda + \mu = 1$ as in 1.22, and we may write

$$f(\lambda x_1 + \mu y_1, \lambda x_2 + \mu y_2, \lambda x_0 + \mu y_0)$$

$$\begin{aligned} &= \lambda^n f(x_1,x_2,x_0) + \frac{\lambda^{n-1}\mu}{1!}\left(y_1 \frac{\partial f}{\partial x_1} + y_2 \frac{\partial f}{\partial x_2} + y_0 \frac{\partial f}{\partial x_0}\right) \\ &+ \frac{\lambda^{n-2}\mu^2}{2!}\left(y_1^2 \frac{\partial^2 f}{\partial x_1^2} + y_2^2 \frac{\partial^2 f}{\partial x_2^2} + y_0^2 \frac{\partial^2 f}{\partial x_0^2} + 2y_1y_2 \frac{\partial^2 f}{\partial x_1 \partial x_2}\right. \\ &\left. + 2y_1y_0 \frac{\partial^2 f}{\partial x_1 \partial x_0} + 2y_2y_0 \frac{\partial^2 f}{\partial x_2 \partial x_0}\right) + \ldots = 0 \end{aligned} \tag{6}$$

since we are assuming that the point (5) lies on $\mathcal{C}_n$.

The equation (6) can be written more compactly if we use the *differential operator*

$$\Delta = y_1 \frac{\partial}{\partial x_1} + y_2 \frac{\partial}{\partial x_2} + y_0 \frac{\partial}{\partial x_0}$$

and the fact that $\Delta^{(n)} f(x_1,x_2,x_0) = n! f(y_1,y_2,y_0)$, so that (6) becomes

$$\lambda^n f(x_1,x_2,x_0) + \frac{\lambda^{n-1}\mu}{1!}\Delta f + \frac{\lambda^{n-2}\mu^2}{2!}\Delta^{(2)} f + \ldots + \mu^n f(y_1,y_2,y_0) = 0 \tag{7}$$

The equation (7) is symmetrical in λ, μ and x, y, and $\Delta^{(2)}$, $\Delta^{(3)}$, etc., denote the *symbolical* square, cube, etc., of the operator Δ; since $\Delta^{(n+1)}$ and all higher powers of Δ annihilate f, we may proceed as before.

By assumption, X lies on $\mathcal{C}_n$ so that $f(x_1,x_2,x_0) = 0$. If we look for the locus of Y such that XY meets $\mathcal{C}_n$ in *two* coincident points $\mu = 0$, we must have $\Delta f = 0$ so that the equation of the *tangent* at X is

$$y_1 \frac{\partial f}{\partial x_1} + y_2 \frac{\partial f}{\partial x_2} + y_0 \frac{\partial f}{\partial x_0} = 0 \tag{8}$$

where $Y(y_1,y_2,y_0)$ is the variable point and $X(x_1,x_2,x_0)$ a fixed point on $\mathcal{C}_n$.

Example 2 Let $f(x_1, x_2, x_0) = x_2x_0^2 - x_1^3$ as in Example 1; then

$$y_1 \frac{\partial f}{\partial x_1} + y_2 \frac{\partial f}{\partial x_2} + y_0 \frac{\partial f}{\partial x_0} \equiv -3y_1x_1^2 + y_2x_0^2 + 2y_0x_2x_0 = 0$$

which reduces to (4) if we set $x_1 = a$, $x_2 = a^3$, and $x_0 = y_0 = 1$.

Let us now suppose that X does *not* lie on $\mathcal{C}_n$ and that the n points of intersection of XY with $\mathcal{C}_n$ are $X_1, X_2, \ldots X_n$. The relation

$$\frac{n}{XY} = \frac{1}{XX_1} + \frac{1}{XX_2} + \ldots + \frac{1}{XX_n} \tag{9}$$

which generalizes 8.43 and defines the point Y for fixed X, can be rewritten in the form

$$\sum_{i=1}^{n}\left(\frac{1}{XY}-\frac{1}{XX_i}\right)=\sum_{i=1}^{n}\frac{YX_i}{XY\cdot XX_i}=0$$

Since $XY \neq 0$ and $YX_i:XX_i = -\lambda_i:\mu_i$, we must have

$$\sum_{i=1}^{n}\frac{\lambda_i}{\mu_i}=0$$

which implies that

$$\Delta f = y_1\frac{\partial f}{\partial x_1}+y_2\frac{\partial f}{\partial x_2}+y_0\frac{\partial f}{\partial x_0}=0 \tag{10}$$

Thus $\Delta f = 0$ *represents the tangent at* X *to* $\mathcal{C}_n$, *if* X *lies on* $\mathcal{C}_n$, *and otherwise it represents the polar line of* X.

We have carried out this same analysis for $n = 2$ in Section 8.4 without using the calculus, but now we see that the method may be generalized and is applicable to any curve $\mathcal{C}_n$ if the polar line of X is defined as in (9) above.

Exercise Prove that if the polar line of X with regard to $\mathcal{C}_n$ passes through Y, then the polar line of Y passes through X.

It is interesting to apply these ideas to determine the condition that a general quadratic polynomial factors into the product of two linear factors. If we take X to be the point of intersection of the two lines represented by $f(x_1,x_2,x_0) = 0$, then every line XY will meet these lines in *two* coincident points at X so that $\mu^2 = 0$; for this to be so we must have $\Delta f = 0$ *identically.* Thus, taking f as in 6.51 (made homogeneous as in 7.243):

$$\begin{aligned}\frac{1}{2}\frac{\partial f}{\partial x_1} &\equiv ax_1+hx_2+gx_0=0\\ \frac{1}{2}\frac{\partial f}{\partial x_2} &\equiv hx_1+bx_2+fx_0=0\\ \frac{1}{2}\frac{\partial f}{\partial x_0} &\equiv gx_1+fx_2+cx_0=0\end{aligned} \tag{11}$$

Since these equations must be consistent, the condition 6.53 follows immediately. More generally, the condition that X be a double point of $\mathcal{C}_n$ is again just that

$$\frac{\partial f}{\partial x_1}=\frac{\partial f}{\partial x_2}=\frac{\partial f}{\partial x_3}=0$$

and the Taylor expansion yields the machinery whereby such double points on $\mathcal{C}_n$ may be studied.

As a final illustration of the power of the method, consider the problem of finding the center of the general conic. In the context of Section

7.5, we saw in Exercises 5 and 6 of Section 8.4 that it is natural to define the center X as the pole of the line at infinity with equation $y_0 = 0$. For (*10*) to so reduce, it is necessary and sufficient that the coordinates of the center satisfy the equations

$$\frac{\partial f}{\partial x_1} = \frac{\partial f}{\partial x_2} = 0 \tag{12}$$

Example 3 In the Example 9.95 the equation 9.951 becomes, in homogeneous coordinates,

$$f \equiv x_1^2 + 2x_1x - x_2^2 - 2x_1x_0 + 2x_2x_0 + 3x_0^2 = 0$$

so that the equations (12) become

$$\frac{1}{2}\frac{\partial f}{\partial x_1} \equiv x_1 + x_2 - x_0 = 0$$

$$\frac{1}{2}\frac{\partial f}{\partial x_2} \equiv x_1 - x_2 + x_0 = 0$$

Setting $x_0 = 1$ we have the nonhomogeneous coordinates of the center to be (0,1), as in 9.959. With this information we can complete the reduction to normal form as previously explained.

It is clear that this approach is applicable to surfaces in space and to more general "varieties," but we shall pursue it no further.

ANSWERS TO EXERCISES

Section 1.2

2. $\frac{1}{\sqrt{3}}, \frac{1}{\sqrt{3}}, \frac{1}{\sqrt{3}}$

3. $x_1 = 1,\ x_2 = \tau,\ x_3 = 2\tau$; no
4. $A'D$: $x_1 = 1 - 2\tau,\ x_2 = 1,\ x_3 = 1$, etc.

5. AB: $\frac{1}{\sqrt{2}}, -\frac{1}{\sqrt{2}}, 0$

Section 1.3

1. Components $(2, -3, 1)$, magnitude $\sqrt{14}$
2. $\overrightarrow{AB}(-3,1,-2)$, $\overrightarrow{BC}(5,-5,-5)$, $\overrightarrow{CA}(-2,4,7)$
3. $|\overrightarrow{AB}| = \sqrt{14}, \left(-\frac{3}{\sqrt{14}}, \frac{1}{\sqrt{14}}, -\frac{2}{\sqrt{14}}\right)$;

   $\left(\mp\frac{3}{\sqrt{14}}, \pm\frac{1}{\sqrt{14}}, \mp\frac{2}{\sqrt{14}}\right)$

Section 1.5

1. $\overrightarrow{AB} = 3i - j - k$
2. $|\overrightarrow{OA} + \overrightarrow{OB}|^2 = 11 = |\overrightarrow{OA} - \overrightarrow{OB}|^2$
3. $|U|^2 = 69, |V|^2 = 65$
4. $\pm\frac{2}{\sqrt{3}}, \pm\frac{2}{\sqrt{3}}, \mp\frac{2}{\sqrt{3}}$
7. (0,0,0)

Section 1.6

2. ABC with equation $x_1 + x_2 + x_3 = 1$ and $A'BC$ with equation $-x_1 + x_2 + x_3 = 1$ intersect at an angle $\cos^{-1}(-\frac{1}{3})$. $AB'C'$ has equation $x_1 - x_2 - x_3 = 1$ and is inclined to ABC at an angle $\cos^{-1}(\frac{1}{3})$. ABC is parallel to $A'B'C'$, etc.
3. $\overrightarrow{AB} = -\boldsymbol{E}_1 + \boldsymbol{E}_2, \quad \overrightarrow{BC'} = -\boldsymbol{E}_2 - \boldsymbol{E}_3$, etc.

Section 2.1

1. 23, 18, 0
2. $x_1 = \frac{7}{23}, x_2 = -\frac{4}{23}, x_3 = -\frac{7}{23}$

Section 2.3

2. 48, 0, 9
3. $x = y = z = 0, w = 2$

Section 2.4

2. $x_1 = 1 - 2k, \quad x_2 = 1 - k, \quad x_3 = 3k$ with direction numbers $(2,1,-3)$; this direction is perpendicular to the normal to $x_1 - 2x_2 = 3$ and so parallel to the plane.
4. $x_1 = 1 + 3k, \quad x_2 = 2k, \quad x_3 = -k$ and
   $x_1 = 3 + 3k', \quad x_2 = 1 + 2k', \quad x_3 = -k'$
5. (i) $x_1 = 3 + k, \quad x_2 = 2 + k, \quad x_3 = k$;
   (ii) $(\frac{1}{2},\frac{5}{2},\frac{3}{2})$

Section 2.5

2. $$\begin{vmatrix} x_1 + 1 & x_2 - 2 & x_3 \\ 2 & -2 & 3 \\ 0 & 0 & -2 \end{vmatrix} = 0 = x_1 + x_2 - 1$$
3. $x_1 + x_3 = 1$

4. $7x_1 + 5x_2 - 3x_3 = 8$

5. $$\begin{vmatrix} x_1 & x_2 & x_3 \\ 0 & 1 & 1 \\ 1 & 0 & 2 \end{vmatrix} = 0$$

Section 3.2

2. $$X^2 = \begin{pmatrix} -5 & 2 & 1 \\ 2 & -2 & 2 \\ 1 & 2 & -5 \end{pmatrix}, \quad XY = \begin{pmatrix} -2 & 4 & -1 \\ -1 & -3 & 1 \\ 4 & 2 & -1 \end{pmatrix}$$

$$YX = \begin{pmatrix} -1 & 1 & -1 \\ 2 & -3 & 4 \\ 4 & -1 & -2 \end{pmatrix}, \quad Y^2 = \begin{pmatrix} -1 & 1 & 1 \\ -2 & -4 & 1 \\ 4 & -2 & -1 \end{pmatrix}$$

4. $X = 0 + X;$

$$Y = \begin{pmatrix} 1 & -\frac{1}{2} & 0 \\ -\frac{1}{2} & 0 & -\frac{1}{2} \\ 0 & -\frac{1}{2} & 1 \end{pmatrix} + \begin{pmatrix} 0 & \frac{3}{2} & 0 \\ -\frac{3}{2} & 0 & \frac{3}{2} \\ 0 & -\frac{3}{2} & 0 \end{pmatrix}$$

Section 3.3

2. $$Y^{-1} = \frac{1}{4}\begin{pmatrix} 2 & -1 & 1 \\ 2 & 1 & -1 \\ 4 & 2 & 2 \end{pmatrix}$$

3. $$A^{-1} = \begin{pmatrix} 1 & 1 & 0 \\ 1 & 1 & 1 \\ 1 & 0 & 1 \end{pmatrix}, \quad (A^t)^{-1} = \begin{pmatrix} 1 & 1 & 1 \\ 1 & 1 & 0 \\ 0 & 1 & 1 \end{pmatrix}$$

Section 3.4

1. 1, 2

Section 3.5

2. $$A^{-1} = \begin{pmatrix} -2 & -3 & -4 & 1 \\ 1 & 0 & 0 & 0 \\ 0 & 1 & 0 & 0 \\ 0 & 0 & 1 & 0 \end{pmatrix},$$

$$B^{-1} = -\frac{1}{16}\begin{pmatrix} -7 & 1 & 3 & -5 \\ -5 & 3 & -7 & 1 \\ 1 & -7 & -5 & 3 \\ 3 & -5 & 1 & -7 \end{pmatrix}$$

Section 3.7

1. $x_1 = -1, x_2 = 0, x_3 = 2, x_4 = 1$
2. (2,3,4,6)
3. (a) (9,17,22,34), (b) (4,1,2,2)

Section 4.2

3. The subgroup I, (12)(34), (13)(24), (14)(23) is normal in $\mathfrak{A}_4$ and also in $\mathfrak{S}_4$. Besides this "four group" the only normal subgroups in $\mathfrak{S}_4$ are I and $\mathfrak{A}_4$, along with $\mathfrak{S}_4$ itself.

Section 4.3

2. (AB) and (AC) lead to the linear transformations

$$Y = \begin{pmatrix} -\frac{1}{2} & \frac{\sqrt{3}}{2} \\ \frac{\sqrt{3}}{2} & \frac{1}{2} \end{pmatrix} X \quad \text{and} \quad Y = \begin{pmatrix} -\frac{1}{2} & -\frac{\sqrt{3}}{2} \\ -\frac{\sqrt{3}}{2} & \frac{1}{2} \end{pmatrix} X$$

Section 5.1

1. (1,0,1,3)
2. $$Y = \tfrac{1}{2}\begin{pmatrix} 0 & 2 & 0 & 0 \\ 1 & -1 & 1 & -1 \\ 1 & -1 & -1 & 1 \\ -1 & 1 & 1 & 1 \end{pmatrix} X$$

4. $\mathbf{X}_2 - \mathbf{X}_3 - \mathbf{X}_4 = 0$
5. $\mathbf{X}_2 - 2\mathbf{X}_1 = 0, \quad \mathbf{X}_4 + 2\mathbf{X}_3 = 0$

Section 5.2

3. $\mathbf{Y}_1 = (1,1,0,0)$, $\mathbf{Y}_2 = (0,0,1,1)$
   $\mathbf{Y}_3 = (\frac{1}{2},-\frac{1}{2},\frac{1}{2},-\frac{1}{2})$, $\mathbf{Y}_4 = (\frac{1}{2},-\frac{1}{2},-\frac{1}{2},\frac{1}{2})$
4. $\mathbf{Y}_1 = \left(\frac{1}{\sqrt{2}}, 0, \frac{1}{\sqrt{2}}\right)$, $\mathbf{Y}_2 = \left(\frac{1}{\sqrt{6}}, -\frac{2}{\sqrt{6}}, -\frac{1}{\sqrt{6}}\right)$,

   $\mathbf{Y}_3 = \left(-\frac{1}{\sqrt{3}}, -\frac{1}{\sqrt{3}}, \frac{1}{\sqrt{3}}\right)$

Section 5.3

1. 4
2. $(1,1,-1)$; $\sqrt{3}$
3. Fourth vertex can be chosen in three different ways; $\sqrt{6}$.
4. 8; $\frac{4}{3}$

Section 5.4

1. $\frac{x_1}{35} = \frac{x_2}{18} = \frac{x_3}{12}$

2. l: $x_1 = 1 + s,\ x_2 = 3s,\ x_3 = 2s$
n: $x_1 = 1 + 2t,\ x_2 = t,\ x_3 = 6 + t$

3. $\dfrac{30}{\sqrt{35}}$

Section 5.5

1. 1
3. $\sqrt{2}$
5. $\frac{1}{6};\ \frac{\sqrt{2}}{6}$

Section 5.8

1. If $\mathbf{Y}_1 = (1,1,0)$, $\mathbf{Y}_2 = (-\frac{1}{2},\frac{1}{2},1)$, $\mathbf{Y}_3 = (1,-1,1)$, then

$$(1,0,1) = \tfrac{1}{2}\mathbf{Y}_1 + \tfrac{1}{3}\mathbf{Y}_2 + \tfrac{2}{3}\mathbf{Y}_3$$

2. $x_1 = t_1 + t_2,\ x_2 = 3t_2,\ x_3 = t_1,\ x_4 = t_2$;
$(\frac{1}{2},3,-\frac{1}{2},1)$, $(1,0,1,0)$

Section 6.1

4. $P(\frac{1}{2},\frac{1}{2})$

5. $$\begin{vmatrix} x_1^2 + x_2^2 & x_1 & x_2 & 1 \\ 1 & 0 & 1 & 1 \\ 1 & 1 & 0 & 1 \\ 13 & -2 & 3 & 1 \end{vmatrix} = 0$$

7. $$\begin{vmatrix} x_1^2 + x_2^2 + x_3^2 & x_1 & x_2 & x_3 \\ -1 & -1 & 0 & 0 \\ -1 & 0 & -1 & 0 \\ -1 & 0 & 0 & -1 \end{vmatrix} = 0; \quad r = \frac{\sqrt{3}}{2}$$

Section 6.2

2. $x_1^2 - 2x_1x_2 + x_2^2 + 2x_1 + 2x_2 - 1 = 0$
5. $x_1^2 + 4x_1x_2 + x_2^2 - 2x_1 - 2x_2 = 0$

Section 6.3

1. $x_1 + x_2 = x_3,\ x_1 - x_2 = 1$
$x_1 + x_2 = 3,\ 3(x_1 - x_2) = x_3$
2. $4x_1 - 2x_2 - x_3 - 3 = 0$

Section 6.4

1. $\frac{x_1^2}{2} + \frac{x_2^2}{3} + \frac{x_3^2}{2} = 1, \frac{x_1^2}{2} + \frac{x_2^2}{3} + \frac{x_3^2}{3} = 1$

   for the first conic, and similarly for the second.

2. (a) $x_2^2 + x_3^2 = 4x_1^2$, (b) $x_2^2 = 4x_1^2 + 4x_3^2$
4. $4a^2(x_1^2 + x_3^2) = (x_1^2 + x_2^2 + x_3^2 + b^2)^2$

Section 6.5

2. All real k; $x_1 - x_2 + 1 = 0 = 2x_1 + x_2 - k$
3. $(x_1 + 2x_2 - 1)(x_1 - x_2 + 1) = 0$
4. $$\mathbf{X}^t \begin{pmatrix} 2 & -1 & 0 \\ -1 & 2 & 0 \\ 0 & 0 & -2 \end{pmatrix} \mathbf{X} = 0$$

   $$\mathbf{X}^t \begin{pmatrix} 4 & -1 & 1 \\ -1 & -2 & 2 \\ 1 & 2 & -2 \end{pmatrix} \mathbf{X} = 0$$

   $$\mathbf{X}^t \begin{pmatrix} 1 & -1 & 2 \\ -1 & 1 & -2 \\ 2 & -2 & 4 \end{pmatrix} \mathbf{X} = 0$$

Section 6.6

1. $x_1 = a \cos \theta \cosh \varphi$, $x_2 = a \sin \theta \cosh \varphi$, $x_3 = c \sinh \varphi$
5. If $|M - \lambda I| = 0$, then $\mathbf{X}^t M\mathbf{X} - \lambda \mathbf{X}^t\mathbf{X} = 0$ represents a pair of planes which meet the quadric $\mathbf{X}^t M\mathbf{X} = 1$ in circles lying on the sphere $\mathbf{X}^t\mathbf{X} = \lambda^{-1}$.

Section 7.2

1. $x_1 = 2t$, $x_2 = t$, $x_0 = 0$
2. $x_1 = x_2 = 0$, $x_0 = 1$; $x_1 = 0$, $x_2 = \pm x_0 = 1$;
   $x_1 = 1$, $x_2 = x_0 = 0$

Section 7.5

5. $$x_1' = \frac{a_{11}x_1 + a_{12}x_2 + a_{10}}{a_{01}x_1 + a_{02}x_2 + a_{00}}$$

   $$x_2' = \frac{a_{21}x_1 + a_{22}x_2 + a_{20}}{a_{01}x_1 + a_{02}x_2 + a_{00}}$$

Section 7.8

2. The internal bisector of $\angle BAC$ is the right bisector of BC, and the external bisector of $\angle BAC$ is parallel to BC.

Section 8.2

1. For $A = \pi/2$: $\cos a = \cos b \cos c = \cot B \cot C$,

$$\frac{1}{\sin a} = \frac{\sin B}{\sin b} = \frac{\sin C}{\sin c}$$

Section 8.3

3. Area $\Delta ABC = \frac{1}{8}$ the area of Σ.

Section 8.4

2. $\frac{x_1x_1'}{a^2} \pm \frac{x_2x_2'}{b^2} = 1$, $x_2x_2' = 2p(x_1 + x_1')$

5. Polar line of (1,0,0) is $x_1 = 0$ and that of (0,1,0) is $x_2 = 0$, and these lines intersect in the center of the ellipse.

Section 9.2

1. $4x_1 - 5x_2 - x_3 = 0$

2. $$\mathbf{X}' = \begin{pmatrix} 1 & 1 & -1 \\ -1 & 1 & 1 \\ 1 & -1 & 1 \end{pmatrix} \mathbf{X}$$

$$\mathbf{X}' = x_1\mathbf{E}_1' + x_2\mathbf{E}_2' + x_3\mathbf{E}_3'$$

Section 9.3

1. $\lambda = 1, 2, 3$, with associated characteristic vectors (1,0,1), (0,1,0), (1,0,−1)
2. (2,3,3), (0,−1,0), (0,−3,−1)

Section 9.4

(a) every point fixed; (b) every point $(a,b,0)$ on $x_3 = 0$ fixed and every line $x_1 + \lambda x_3 = 0$ fixed; (c) (1,0,0) fixed and $x_3 = 0$ fixed; (d) for $a \neq 1$, every point $(a,b,0)$ on $x_3 = 0$ fixed and also (0,0,1) fixed, so that every line through (0,0,1) is fixed; (e) for $a \neq 1$, (1,0,0) and (0,0,1) fixed so that $x_2 = 0$ is fixed.

Section 9.5

1. 3 in each case.
2. The rank is unchanged.
3. 4 in each case.
4. (a) 2, (b) 2, (c) 1

Section 9.6

1. $$P = \begin{pmatrix} \frac{1}{2} & 0 & \frac{1}{2} \\ 0 & 1 & 0 \\ \frac{1}{2} & 0 & -\frac{1}{2} \end{pmatrix}, \quad P^{-1} = \begin{pmatrix} 1 & 0 & 1 \\ 0 & 1 & 0 \\ 1 & 0 & -1 \end{pmatrix}; \quad \text{no}$$

2. (b) $\begin{pmatrix} a & 0 & b \\ c & d & e \\ 0 & 0 & d \end{pmatrix}$

Section 9.7

1. $(\frac{1}{3},\frac{2}{3},\frac{2}{3})$, $(\frac{2}{3},\frac{1}{3},-\frac{2}{3})$, $(\frac{2}{3},-\frac{2}{3},\frac{1}{3})$

Section 9.9

2. (a) $(x_1 - x_2 - 1)(x_1 - 2x_2 + 1) = 0$
   (b) $2(x_1 + x_2)^2 = (x_1 - x_2 - 1)$
   (c) $(x_1 + x_2)^2 + 2(x_1 - x_2 - 1)^2 = 1$
   (d) $2(x_1 + x_2)^2 - (x_1 - x_2 - 1)^2 = 1$
   (e) $2(x_1 - \frac{1}{2})^2 + 2(x_2 + \frac{1}{2})^2 = -1$

Index

A

B

C